UNDERSTANDING
CHEMISTRY
THROUGH CARS

Understanding
CHEMISTRY
THROUGH CARS

Geoffrey M. Bowers
Ruth A. Bowers

CRC Press
Taylor & Francis Group
Boca Raton London New York

CRC Press is an imprint of the
Taylor & Francis Group, an **informa** business

CRC Press
Taylor & Francis Group
6000 Broken Sound Parkway NW, Suite 300
Boca Raton, FL 33487-2742

© 2015 by Taylor & Francis Group, LLC
CRC Press is an imprint of Taylor & Francis Group, an Informa business

No claim to original U.S. Government works

Printed on acid-free paper
Version Date: 20140715

International Standard Book Number-13: 978-1-4665-7183-9 (Paperback)

This book contains information obtained from authentic and highly regarded sources. Reasonable efforts have been made to publish reliable data and information, but the author and publisher cannot assume responsibility for the validity of all materials or the consequences of their use. The authors and publishers have attempted to trace the copyright holders of all material reproduced in this publication and apologize to copyright holders if permission to publish in this form has not been obtained. If any copyright material has not been acknowledged please write and let us know so we may rectify in any future reprint.

Library of Congress Cataloging-in-Publication Data

Bowers, Geoffrey M. (Geoffrey Mark)
   Understanding chemistry through cars / Geoffrey M. Bowers, Ruth A. Bowers.
     pages cm
   "A CRC title."
   Includes bibliographical references and index.
   ISBN 978-1-4665-7183-9 (pbk. : alk. paper) 1. Chemistry, Technical--Popular works.
   2. Chemistry, Technical--Study and teaching. 3. Automobiles--Equipment and
   supplies--Study and teaching. I. Bowers, Ruth A. II. Title.

TP147.B69 2015
629.222--dc23
                                              2014027281

Visit the Taylor & Francis Web site at
http://www.taylorandfrancis.com

and the CRC Press Web site at
http://www.crcpress.com

Contents

Preface

The journey to create this book began on a long walk to our office during graduate school in 2003. Walking from the stadium parking lot at Penn State to the chemistry buildings gave us plenty of time to think and talk, and one morning we were particularly excited about brainstorming ways that chemistry can be used to describe cars. Immediately upon arriving at work, we typed up our ideas, thinking that some day we would teach a summer camp or course with this theme. Occasionally, we would add to the list and daydream ways to present the material. Nine years later, while our eldest son was napping during a daylong drive to visit family for the holidays, we developed a draft table of contents and decided to write up our ideas as a book.

We put the time and effort into writing this book because we learned so much from one another simply while brainstorming the topics, and we were astounded that there was no such text already. Geoff's inter-disciplinary chemistry and engineering background and enthusiasm for cars, combined with Ruth's interests in education and the chemistry of consumer products, gave us a unique perspective on marrying and marketing the "chemistry of cars." We both firmly believe in teaching chemistry by stressing practical applications, as we want our students to become informed consumers of everyday chemistry who can make educated decisions on personal, environmental, health, and political matters. The automobile is a great example of a consumer product with an abundance of chemistry hiding in plain sight, and personal transportation has an impact on a wide variety of socially important issues, including energy and the global carbon balance. There is almost nothing in an auto-mobile that cannot be described from a chemical perspective, but for this book, we have chosen several major car systems and chemistry concepts to showcase the links between our passions.

As a reader, we expect that you picked up this book already having some level of interest in either chemistry or cars, with a desire to learn more about both. We have provided some basic chemical and automotive background in the appendices, so you can brush up on general chemistry topics or the basic systems in a car if you feel the need. Some of the chemistry topics presented in the book do build on one another, so the chapters are best read in order if you do not have a strong background in chemistry already. If you are using this book in a chemistry course or in conjunction with a general chemistry textbook, we hope you can skip around to meet your course goals. As you are reading, you may have a desire for information beyond what we feel is within the scope of this book. We have provided many references that guided us for specific topics, but you may find more information in sources as varied as patents, automotive technician manuals, chemistry textbooks, and handbooks on the chemistry of specific classes of compounds (such as pollutants or polymers). Additionally, there are many diagrams, animations, and videos available from various online resources that may help you better visualize concepts described in this text. We encourage you to look up a topic or phrase online and to use these preexisting visualization tools as often as possible. Finally, we love teaching and we are passionate about cars, so we are very interested in interacting with you if you are reading this book. To that end, we have made ourselves available to you the readers via a dedicated Twitter account (@CarChemProf) and an associated blog (http://www.thechemistryofcars.com/). Please visit these to ask us direct questions and see responses from us and other experts or to hear the latest news about car chemistry.

Despite our driving passions, this book would not be what it is without a great deal of support from many individuals. Many thanks go to several people at Taylor and Francis, including Lance Wobus for taking a chance on our idea, Barbara Glunn our editor, and to David Fausel for coordinating the project. Thanks also go to Dr. John D'Angelo for introducing us to Lance when we first decided to write this book. Several individuals also graciously reviewed the book draft one or even several times. We extend our thanks to Larry Bowers, Peter Wagoner, Arielle Polakos, and David Bish for helpful comments and suggestions that helped develop the book into what it is today. Thanks are also due to Geoff's parents, Larry and Janet Bowers, for providing us with a calm and peaceful working environment to bring the book draft near completion during the summer of 2013. Geoff would also like to thank Larry for instilling both his love of cars/racing and his interests in chemistry—there would be no book without your influence over the years. In addition, we thank Ruth's parents, Bret and Peggy Rivers, who were probably surprised when we arrived at their home for a visit and immediately requested paper and pen to jot down our initial outline for this text. Peggy's visit, cooking, and grandson wrangling during a spring break from school were also essential in the final stages of revision.

This book is dedicated to Isaac and James. May our love of cars be passed down to you and give us all many years of joy and togetherness.

About the authors

Geoffrey M. Bowers is an interdisciplinary researcher and educator who has been teaching chemistry to undergraduates and supervising graduate and undergraduate research at small liberal arts colleges since 2008. Geoff holds a Bachelor of Science degree in chemical engineering and a Cooperative Education Certificate from Purdue University as well as a PhD in chemistry from The Pennsylvania State University. After earning his  PhD for studies of strontium binding in natural minerals using ^{87}Sr nuclear magnetic resonance (NMR) spectroscopy, Dr. Bowers spent several years as a postdoctoral research associate at the University of Illinois in the Department of Geology and Michigan State University in the Department of Chemistry, where he furthered his expertise in NMR by using the technique to study both structural and dynamical behavior at mineral–fluid interfaces. He completed one year of his post-doc while also serving as a visiting faculty member in the Chemistry Department at Gustavus Adolphus College teaching general and environmental chemistry. Geoff is currently an assistant professor of chemistry and an adjunct assistant professor of materials engineering at Alfred University, a regional university in southern New York focused on undergraduate education. Since arriving at Alfred, Bowers has taught a variety of courses including general chemistry, thermodynamics, kinetics, quantum mechanics, environmental chemistry, and several topics courses including "The Chemistry and History of the Manhattan Project" and "The Chemistry of Cars." Dr. Bowers's educational philosophy emphasizes the value of and advocates for student-faculty co-learning experiences (such as undergraduate research), employs a variety of student-centered techniques in the classroom, uses real-world examples whenever possible, and stresses authentic assessment and teamwork skills in his courses. To keep up with educational best practices, Geoff is active in the Project Kaleidoscope Upstate New York Regional Network. He recently completed an 18-month term as a co-leader of the network steering committee, which he still serves on as a delegate. Over his career, Dr. Bowers has been the recipient of several teaching awards and an undergraduate research mentor award. In addition to his professional academic history, Dr. Bowers has two years of industrial R&D experience in the composite wet friction material industry, where he worked on engineering composite materials for heavy-duty and automotive clutch plates.

Ruth Bowers is a chemistry educator specializing in teaching chemistry at the high school/collegiate interface. She earned a BS in chemistry from Purdue University with an environmental chemistry specialization. During her time at Purdue, Ruth became interested in chemistry education while working as a tutor in the chemistry resource room and collaborating with the chemistry education faculty about curricular design. Ruth then attended graduate school at The Pennsylvania State University where she completed work in both the chemistry and education departments, ultimately earning her MEd in curriculum and instruction and a Pennsylvania state teaching certification. For her master's degree, she studied the impact of an "atoms first" approach to general chemistry (where the course begins with atomic structure and builds in size up to reactions steadily over the semester) on student comprehension of select chemistry topics. Ruth also received a teaching award from the Department of Chemistry for being an outstanding laboratory teaching assistant and served as a trainer and supervisor to both laboratory and recitation teaching assistants. Since then, Ruth has served as an adjunct instructor at Penn State University, Gustavus Adolphus College, Alfred State College, and Alfred University, teaching general chemistry laboratories and lecture courses. In her current position as adjunct instructor of chemistry at Alfred University, Ruth also developed and implemented curricula for teaching the chemistry of consumer products as an honors seminar course and a regularly occurring high school summer camp.

Geoff is an automotive enthusiast who loves working on cars, driving, and watching many types of racing. He has been to many major auto shows in the United States and several types of racing events including the Indianapolis 500, the United States Grand Prix (Formula 1), the NHRA US Nationals, Pikes Peak Hill Climb, Grand American Six Hours of Watkins Glen, and others. Ruth also has ties to automotive history, as she spent her childhood mere miles from the site of one of the first successful test drives of a gasoline-powered automobile in the United States. When not focused on racing or teaching, Geoff and Ruth both enjoy spending time with their sons Isaac and James and their basenji Cooper, gardening, skiing, playing hockey, fishing, and being outdoors with family, friends, and their camera.

Introduction: Cars and chemistry

The car

Cars have captured the imagination and stirred the passion of millions of men and women worldwide since the first car appeared in 1886. Iconic carmakers like Porsche, Ferrari, and BMW are now recognized across the globe, as are timeless iconic cars like the Corvette, Shelby Cobra, Porsche 911, and Ferrari F40. Whether you love luxury and refinement, raw muscle-car power, protecting the environment, or the ultimate blend of art, science, and engineering that is the exotic super car or race car—there is an automobile club, and network of supporters/suppliers out there that share your interests. Many men and women gladly volunteer their weekends to work as marshals at various types of local to international racing events that span the four seasons (SCCA, rally, drift, driving school, etc.); automotive enthusiasts can be found at car shows, car auctions, drag strips, and racetracks year-round; and every city and town in America seems to have at least one automobile dealer, tire dealer, auto parts store, and/or mechanic's shop. You would be hard pressed not to find several television shows about automobiles on the air every day, and cars at one point even spawned their own 24-hour network (Speed). Truly, there is a global subculture of people who dedicate their lives to learning about, maintaining, modifying, racing, and loving cars.

Why are people so fascinated with cars? We challenge you to attend a local SCCA, NHRA, IMSA/Sports Car, NASCAR, or Indy Car rally or other racing event and interact with the drivers, mechanics, and fans without taking away some level of passion about cars. The smooth and sharp lines of a race car are not only aesthetically pleasing, but all are placed with a great deal of care and precision to provide an engineering function and/or racing advantage. The car demands your wonder because it is a conundrum of the simple and astoundingly complex. For example, the beautifully straightforward yet intricate nature of an engine and the sound of it roaring to life can put smiles on faces of any age and education. You can almost feel the car itself anticipating its dance around the racetrack or down the drag strip as the engine growls and pops and all your senses respond to becoming immersed in the exhaust vapors, smells, and sounds of raw speed and excitement. Backyard mechanics get the satisfaction of doing something themselves—and doing it right—when they repair or modify a car. Young and old find a medium for expressing themselves by building the outrageous, the sleeper, the one-of-a-kind paint job, or even covering the windows and tailgate with loud stickers expressing their passions. Working on cars and appreciating cars with family is a great way to spend time together and pass on a legacy to the next generation, as well as saving or even making money. I find myself struggling to keep my heart rate down and not burst from excitement just writing this section of the book.

Yet with all the moving power of a car and the hundreds of millions who use them every day, we suspect many would have a hard time answering the question, "What is a car?" It is a machine clad in artwork harnessing the power of explosive chemistry. It is a network of interconnecting systems all working together to move thousands of pounds of man and machine toward a goal. A car is a collection of parts that are the brainchild of thousands of people and more than 100 years of innovation. It is a gateway to freedom, a means of expression, or just a way of getting from here to there. And it is also exceedingly dependent on chemistry—for the energy to move, the tires that hug the road, the polymers that form the interior, the variety of fluids that permeate the many systems, the colors in the paint, and the headlamps that light your way at night.

Chemistry

Chemistry is the study of matter and the changes it undergoes. More specifically, chemistry involves the study of matter with consideration of atoms and molecules, the interaction of different types of matter together, and the interaction of matter and energy. Chemists are interested in everything from reactions to physical properties and study materials from biomolecules to terrestrial and extraterrestrial rocks and soils. Today, chemistry is often called the *central science* because it provides a bridge between the physical, life, and earth sciences, as well as connecting these fields with health and engineering disciplines.

Chemistry has its ancient roots in alchemy, a study that blended early science with philosophy. While many goals of alchemy were unachievable, such as turning common metals into precious ones, the methodical study and documentation of the work by alchemists caused them to shape our ways of understanding the natural world. In addition to developing basic chemical methods such as ore extraction and distillation, alchemists created a framework to understand and organize matter.

Alchemy eventually gave way to the branches of science we know today because advances in mathematics and analytical methods permitted more in-depth study of the natural world. In the late seventeenth through early nineteenth centuries, work by Robert Boyle (characteristics of gases), Antoine LeVoisier (conservation of mass), and John Dalton (atomic theory) defined the new study of chemistry and set the stage for what chemistry has become today. These early chemists differed from alchemists in that their goals were to develop theories to describe the world, rather than creating or transforming substances.

With the development of atomic theory and a better understanding of the nature of matter came a need to organize the known elements and a search for commonality between chemical species. In the mid– to late–nineteenth century, Dmitri Mendeleev published a periodic table that not only systematically organized known elements, but predicted the existence of several others that would be discovered later in time, thereby securing its credibility. Since that time, the periodic table has undergone some minor organizational changes, but most chemistry students can still recognize the original pattern described by Mendeleev.

Modern chemistry has come a long way from the roots of alchemy, but there are still many exciting challenges in need of bright new chemists to resolve. A few notable discoveries in modern chemistry and chemical physics include the creation of synthetic elements via radiochemical methods, recognition, and implementation of catalytic methods for achieving chemical reactions, development of the quantum mechanical theory of the atom, the discovery and advancement of computational chemistry, and the astounding advances in analytical methods. Today, chemists play crucial roles in addressing many socially important questions, from developing methods to duplicate rare and valuable products available in nature, to improving sensitivity and developing new analytical techniques to improve animal or environmental health, to discovering more environmentally friendly methods for making existing and new chemical products, to inventing new methods for harvesting and storing energy, and to managing our global greenhouse gases. Chemists now work in one or more of the traditional subdisciplines (organic, inorganic, analytical, or physical) or combine their studies with another science to work in an interdisciplinary field. They are often part of large collaborations of scientists from varied disciplines and backgrounds working on a common problem from different approaches. Chemists are also involved in automotive pigments and coatings, the tire and rubber industry, the glass science community, and many other areas/groups whose work is integral to cars and their production.

Chemistry in cars

Whether you are an auto enthusiast wanting to know more about the science behind your favorite ride, a chemist wanting to know more about the parts under the hood of your transportation to work, or a student interested in practical applications of science, we think that you will be able to find something of interest about both chemistry and cars in this book.

We start in Chapter 1 with a discussion of gases, one of the earliest interests in modern chemistry and one of the first ideas that prompted us to write this book. Gases are essential inputs and outputs of a combustion engine, and their properties make it possible for us to have comfortable suspensions, inflated tires, cushioning foams, air bags, and climate control. One of the great things about gases is that you probably already have a basic understanding of the fundamental behavior of gases developed through your own experiences. Knowing that a hot-air balloon rises because its air is less dense than the atmosphere translates into knowing that a cold-air intake on a car can increase the amount of oxygen available for combustion in your car's engine.

In Chapter 2, we delve into the fundamental reaction that occurs in every internal combustion engine: hydrocarbon combustion. Building on the gas laws, we can describe the physical and chemical processes that occur in four-stroke gasoline and diesel engines. Additionally, we discuss ways of describing the complex fuel mixture known as gasoline, the energy released during combustion, the ways to improve engine efficiency and power (such as turbochargers and superchargers), and alternatives to petroleum-derived fuels.

Chapter 3 continues our discussion of combustion and brings to light a series of other important automotive reactions called oxidation-reduction (REDOX) reactions. Combustion itself is a REDOX reaction that is revisited here, and we explore other reactions, including the basis for the operation of a battery, catalytic converters, and the dreaded rusting of a classic car. Current hybrid electric vehicles would not be possible without past advances in battery chemistry, and research in the fields of battery improvement and hydrogen fuel cell technology are going strong today. Both require significant knowledge of REDOX and materials chemistry.

Intermolecular forces, the interactions that attract similar molecules together, are at the heart of Chapter 4. As in the case of gas laws, you probably have many prior experiences that prepare you for understanding this topic. Just as oil and vinegar (an aqueous organic acid) separate in salad dressing, wax and other organic coatings repel water from your car. Intermolecular forces are also essential to describe the behavior and value of such diverse products as engine oil, gasoline additives, and detergents.

In Chapter 5, we build upon the energetic concepts introduced in our discussion of combustion and discuss ways of managing the temperature of critical automobile components and systems, including you, the driver. This chapter combines ideas from previous chapters, such as energy, thermochemistry, solutions, and intermolecular forces.

Chapter 6 covers the materials chemistry of the car, which helps us understand various components such as plastics, rubbers, composites, and alloys. We discuss polymers and rubber chemistry that build upon the elements of organic chemistry presented earlier and present a more detailed look at the what and why of automotive alloys. We end this chapter with a discussion of hydrogen fuel cells and some of the chemistry roadblocks that must be overcome before vehicles powered by these devices can become a safe and reliable mode of transportation.

Chapter 7 deals with the ways that light interacts with your car. We discuss how light of various energies interacts with the molecules of your car, and we use these ideas to explore paints and the light-based degradation of plastics and other polymers. We also cover the light that your car generates in headlamps and taillights. You have probably seen advances in domestic light bulbs in recent years, and car headlights and taillights have benefited from this new technology to generate clearer, brighter, and longer lasting interior and exterior lighting.

Perhaps our greatest motivation for writing this book is that the car represents an ideal and untapped medium for engaging with, exploring, and building excitement about the science of chemistry to STEM-focused* high school, technical school, and college students of all ages. For example, cars can be used to tie together the material presented in a typical undergraduate chemistry curriculum, making car chemistry an ideal topic for a synthesis-type (as defined in Bloom's taxonomy) college capstone course. Faculty and students interested in this avenue are encouraged to use this book as a foundation for developing broader questions about car chemistry and/or to explore topics in this book independently and in greater detail, leading to more complex calculations, discussion of more intricate reactions and reaction mechanisms, etc. Ideas for helping make this book function in that setting can be found in Appendix E. At the same time, cars have great general appeal and can easily be used to teach chemistry to upper-level high-school students and beginning college students and inspire them/you to pursue chemistry or STEM-related goals. From these perspectives, the book is intended either as the core material for a collegiate outreach summer program, as a source of real-world examples for college faculty to use in the classroom, or as a topic of personal interest.

The checkered flag
Our primary objectives in this text are to help the novice car user understand cars through chemical principles, to help teach the novice chemist basic chemical principles and their value in the context of cars, and to help upper-level chemists synthesize the knowledge they have gained and examine one way it manifests in the real world. A secondary goal of

---

* Science, technology, engineering, and mathematics.

the book is to increase public interest in the chemical sciences, and a third is to introduce a deeper understanding of cars and a passion for them in our readers. We hope that when you finish the book, you will be able to describe and discuss a car in much more detail than you could before reading the text. At the same time, we expect that you will be able to intelligently discuss chemical issues related to the car, perhaps using cars as a means to engage others about the exciting science of chemistry. More specific learning outcomes have been provided at the beginning of each section to help you monitor your own learning progress. We also hope that when you finish the text, you will have developed a deeper appreciation for both chemistry and your car. Enjoy the book!

chapter one

The properties and behavior of gases

Gases and gas-phase chemistry play very important roles in the world around us and critical roles in several important systems in cars. Gases are often used in shock absorbers and struts, are released during combustion reactions, are the basis of power generation in internal combustion engines, and are critical to the efficient operation of these engines from several perspectives. In this chapter, we will introduce some of the basic concepts used to understand gas behavior and apply these concepts to gain a greater understanding of the automobile. Since gases are so pervasive in the structure, function, and operation of cars, many of these concepts will be revisited in subsequent chapters.

1.1 Kinetic molecular theory (KMT)

Chemistry Concepts: thermodynamics, statistics, gas laws
Expected Learning Outcomes:
- Explain the basic principles of KMT
- Identify the inaccurate assumptions of KMT

Gases are the least dense and most compressible state of matter. Because their density is so low, we often consider gas behavior without worrying about interactions between molecules. These intermolecular interactions significantly complicate our understanding of chemistry in liquids and solids. However, the low density of gases and the subsequent minor role of intermolecular forces in gas behavior allow us to gain significant insight about gases with relatively simple and straightforward models like the ideal-gas equation or the van der Waals gas equation. These models, particularly the ideal-gas equation, are based upon a well-established theory about gas behavior known as kinetic molecular theory (KMT).

Kinetic molecular theory provides a physical rationale that explains the empirical relationships (those observed in the laboratory) between gas temperature, pressure, volume, number of molecules, etc. It also helps to explain one of the major reaction theories in chemistry known as collision theory, which says that reactions occur when molecules collide with the appropriate energy and in an appropriate orientation. Kinetic molecular theory is relatively simple, and it is actually quite remarkable that five basic statements involving several unrealistic assumptions about gas behavior can make sense of well-known relationships between gas properties. Kinetic molecular theory says:

1. A gas is composed of molecules separated by distances much greater than the size of the molecules themselves. Because of these large separations, the gas molecules can be considered as points in space that take up no volume.
2. Gas molecules are in constant motion in random directions.
3. Gas molecules frequently collide, and all collisions are elastic (kinetic energy is conserved).
4. Gas molecules exert neither attractive nor repulsive forces on one another.
5. The average kinetic energy of a collection of gas molecules is proportional to the temperature in Kelvin (K). Any two gases at the same temperature will have the same average kinetic energy.

Kinetic molecular theory leads directly to the derivation of the ideal-gas equation (see Section 1.5), which is useful for understanding many gas phenomena, including the roles of gases in several fundamental aspects of the automobile.

Let us examine several aspects of kinetic molecular theory in greater detail. Both the first and fourth points are related to our earlier discussion of gas density. In fact, the statement "separated by distances much greater than the size of the molecules themselves" is actually a fancy way of saying "a low-density phase." In KMT, the low density is manifested not only as point four, which states that intermolecular forces are ignored in this model, but also that the volume of the gas particles themselves is ignored, which is the second half of point one. If the distance between particles is many orders of magnitude larger than the size of the particles themselves, then the volume occupied by the gas particles is very small. As we approach an infinite separation between the particles, whatever volume they truly occupy becomes negligible. The "kinetic" part of KMT comes from the second, third, and fifth statements. It seems logical that the particles will be in constant motion, since they have a very low mass and they would be a solid or liquid if the molecules were moving more slowly. The elasticity of collisions, item three, simply means that kinetic energy is not transformed into potential energy or any other form during collisions between gas molecules. Two molecules in an arbitrary collision may have different speeds before and after the collision, but the sum of their kinetic energies before and after will be identical. This also relates to the key point hidden in statement number five, namely that there is a distribution of molecular speeds in a gas. In the room you are in, there is a likelihood that some gas molecule is barely moving and that some molecule is moving so fast that it hits each boundary in the room several times while you read this sentence. But there are an enormous number of gas molecules in any room. In fact, the number is so large that it is impossible to measure the kinetic energy of any individual particle at any point in time (for reference, a 215/55R16 tire holds $\approx 3.5 \times 10^{24}$ gas molecules when filled to a typical internal pressure of 35 psi at 25°C).

So how do we estimate the kinetic energy of gas particles? It turns out that, for gases, temperature is actually a measure of the average kinetic energy. We can't measure the temperature or kinetic energy of an individual molecule, though. A thermometer placed in a room is struck by a large number of gas particles in any given interval of time, meaning that temperature always gives the average kinetic energy of the gas molecules in a space. In fact, some relatively simple calculations reveal that at 273 K (0°C) and a pressure equivalent to 1 atm, nitrogen gas molecules experience on average $\approx 10^{10}$ (10 billion) collisions per second. It is also important in kinetic molecular theory to remember the difference between velocity and kinetic energy. The distribution of molecular velocities will change as the mass of the gas particles varies (recall that kinetic energy of a moving body is proportional to the mass times the velocity squared), but the fifth point indicates that the total kinetic energy is always directly proportional to the temperature. Essentially, a heavy body and a light body at high temperature have the same high kinetic energy; however, the heavy body will be moving with lower velocity, since it possesses greater mass. This fifth point also illustrates one of the basic tenets of statistical mechanics, that the number of entities in a chemical system is always so large that we observe weighted averages when we measure physical properties.

As noted earlier, many of the statements in kinetic molecular theory involve assumptions that are physically unrealistic. For example, gas molecules do have a real size and take up some of the volume in a container, in contrast to statement one. It is also unlikely that every collision between gas molecules is elastic. Some energy may be converted to another form, such as molecular rotation or vibration rather than translational energy. Statement four is certainly not true in the case of large molecules or high gas densities. For example, large noble gas molecules can be observed to pair in molecular dynamics simulations of gas behavior, presumably due to strong dispersion forces, which are an intermolecular attraction that is directly proportional to the size of the molecule (see Chapter 4). However, removing many of the assumptions that seem unreasonable actually cause very small changes in the relationships predicted by the ideal-gas equation and kinetic molecular theory. For example, the van der Waals equation eliminates the assumption that the gas particles have no volume with a volume correction term and acknowledges molecular interactions in an interaction correction, both of which involve empirically determined constants. In both cases, these corrections produce significant variations only for large gas molecules and high gas densities, meaning the ideal-gas equation is more than adequate under many conditions.

While kinetic molecular theory may seem to have little to do with a car, it actually explains the source of power in conventional internal combustion engines. As we will discuss in Chapter 2, burning a fossil fuel converts the chemical energy stored within the chemical structure to kinetic energy of the gas molecules and what is termed *work of expansion*. It is partly the kinetic energy of the gas-phase molecules that pushes on the piston, turning the crankshaft, and ultimately turning the wheels. The gas laws and KMT are the heart that drives the internal combustion engine.

1.2   *Tires, pressure, and pressure units*

Chemistry Concepts: dimensional analysis, kinetic molecular theory, statistical thermodynamics

Expected Learning Outcomes:
- Understand how gas pressure is related to the motion of gas-phase molecules
- Define pressure and use dimensional analysis to convert between common units
- Use tire pressure to explain the difference between gauge and absolute pressure measurements
- Use the behavior of tires to prove that a relationship exists between temperature and pressure of gases

Have you ever wondered why you need to add air to your tires in the winter or release air in the summer to maintain ideal tire pressure? Or have you observed that the pressure in your tires is higher after you go for a drive than before you left? Or heard race drivers and teams talking over a pit radio about warming up tires on race day? All of these behaviors relate directly to the concept of pressure, specifically the relationship between temperature and pressure that is implied in kinetic molecular theory, which is the subject of this section.

To begin, we must define pressure, which is the amount of force exerted per unit area. You can experience pressure yourself simply by standing flat on your feet, then standing on your tiptoes. The pull on your body due to gravity hasn't changed, but you are concentrating that force over different areas. The smaller area supporting your body weight when on your tiptoes increases the pressure on your toes, and the pain signal sent by your nerves to your brain are a direct indicator of that change in pressure. In the case of gas-rich systems such as our atmosphere, the force generating pressure arises from collisions between the gas molecules and the boundaries of an object rather than the downward pull of gravity in our tiptoe example. One can easily observe that gas pressure does not have a direction. In fact, if it did and gravity were solely responsible for gas pressure, we would likely all be flat due to the weight of the atmosphere crushing down upon us. This nondirectionality of gas pressure relates back to kinetic molecular theory and the fact that particles are constantly in motion in random directions, meaning that the likelihood of being hit by a particle on the side of your foot is the same as being hit by a particle directly on top of your head. Statistical mechanics tells us that when there are a large number of particles, as there always are in macroscopic chemical systems, a large number of collisions occur at every point on a surface in any one observable interval of time, supporting the idea that pressure is relatively constant in all directions.

Kinetic molecular theory and the definition of pressure also tell us that there must be some sort of relationship between gas pressure and gas temperature. We know that force is mass times acceleration, and it is known that acceleration is the derivative of velocity. From Section 1.1, we know that kinetic energy is also proportional to mass and velocity (velocity squared, to be precise). Thus, pressure and kinetic energy both depend in some way on mass and velocity, and if KMT tells us that temperature is a measure of the average kinetic energy in a system of gas particles, there must be some link between gas pressure and temperature. This has been demonstrated in the laboratory and is known as Gay-Lussac's law, named for Joseph Gay-Lussac, who first observed the effect in 1809.

$$\frac{P_1}{T_1} = \frac{P_2}{T_2}$$

It is imperative that in Gay-Lussac's law (and all of the other gas laws involving temperature) that you use absolute temperature in Kelvin (K) when you perform a calculation. Gay-Lussac's law is the first of many "simple" gas laws we will explore in the rest of this chapter, and it explains quite a bit about tires and tire pressures. For example, "warming up" the tires in a race car is in part directly related to Gay-Lussac's law, as we will see in the following discussion.

There are many different units that can be used to discuss pressure, and the most appropriate unit is often determined by the specific situation under consideration. For example, if we are discussing pressures deep inside the Earth, the pressure is so high that it would not be very useful to use a pressure unit with a small increment such as pounds per square inch (psi). Use of psi in geological systems would necessitate the use of very large numbers that are difficult to interpret, and there is no need for that type of precision. Shallow geological pressures are often better considered in kilobars, or kbar (1 kbar is roughly 1000 atm, or equivalent to 1000 times the pressure exerted by our atmosphere at 273 K at sea level).

In the case of cars, the pressures we are working with are rarely more than a few atmospheres, which make a pressure unit with a small increment like psi a reasonable choice due to the finer precision with which pressure can be measured. The common units of pressure and their equivalencies with respect to atmospheric pressure are listed in Table 1.1.

A discussion of pressure-measuring devices can easily be found on the Internet or in a general chemistry or general physics textbook. However, an issue related to pressure measurement that is less commonly discussed in chemistry is whether the pressure due to the atmosphere is included in the measurement or not (Figure 1.1). Clearly, if a tire indicates it is to be inflated to 35 psi, you need to know if this is 35 psi beyond atmospheric pressure or if it is a total of 35 psi in the tire, and you need to know whether your pressure gauge accounts for atmospheric pressure or not. If you make a mistake, you could end up

Table 1.1 Common Pressure Units, their Symbols, and Equivalencies

Pressure unit	Symbol	Equivalence (with respect to 1 atm)
Atmosphere	atm	1
Torr	torr	760
Millimeters of mercury	mm Hg	760
Bar	bar	1.01
Kilopascals	kPa	101.3
Pounds per square inch	psi	14.7

Figure 1.1 Difference between gauge (top) and absolute (bottom) pressure.

over- or underinflating your tire by about 50%! Underinflation reduces fuel economy and increases the chance that the tire will burst under normal driving conditions. In chemistry, physics, and engineering, *absolute pressures* are those that include the pressure due to the atmosphere in their value. For example, to fill a tire to 35 psia (psi absolute), you will need to add 35 psi – 14.7 psi = 20.3 psi to the tire. *Gauge pressures* are those that exclude the contribution due to the atmosphere, and it is gauge pressure that we commonly find written on the side of tires. When you place a tire on a wheel, the tire already holds gas from the atmosphere. As we now know, those gas particles should have the same kinetic energy as those in the atmosphere and collide with the tire walls in all directions, i.e., the tire must have close to 14.7 psi of pressure already. However, that tire is considered "flat," and the gas inside the tire will not support much (if any) of the weight of the vehicle. If you fill that tire to 35 psig (psi gauge), it will really contain 14.7 psi + 35 psi = 49.7 psi of total pressure, and the gas inside the tire will nicely support the weight of the vehicle. Nearly all tire gauges read gauge pressure. You can verify that your pressure gauge reads gauge or absolute pressure by noting the pressure before placing it on the tire stem: If it reads zero, the device measures gauge pressure. If it reads a value between 14 and 15 psi, it displays absolute pressure. It is very important when using the simple and combined ideal-gas laws that you use absolute pressures in your calculations.

Car tires represent an ideal example of the pressure/temperature relationship revealed in Gay-Lussac's law. In racing, drivers often turn their cars violently back and forth across the track behind the pace car before a green flag is waived. They also allow gaps to build between the car they are pursuing before the green flag, then slam on the gas and spin the drive wheels. If you are a fan of drag racing, you know that spinning the drive wheels before a race occurs routinely, with Pro-Stock, Funny Cars, and Top Fuel Dragsters all beginning the race with a significant burnout at the starting line, frictionally heating the drive tires via significant wheel spin. Formula-1 cars can often be observed in the paddock before qualifying or before a race with electric tire-warmers wrapping the wheels. In all of these cases, the drivers/teams are warming up their tires. Part of the reason for these preracing actions is surely to optimize the level of grip, or the coefficient of friction between the tire itself and the road surface, which is related to the "tackiness" of the rubber and the rubber chemistry as well as mechanical properties of the tire rubber. But some of it is also related to tire pressure. Let's say that your team filled your tires to 40 psia the night before a race when the temperature was 15°C (288 K). They move the tires out to the pit box in the morning, where they sit in the hot sun until they go on your car during the race. Perhaps the tires warm to 30°C (303 K) in the sun. What is the pressure in the tires now? Using Gay-Lussac's law, we see that the final pressure is now 42.1 psia:

$$\frac{p_1}{T_1} = \frac{p_2}{T_2}$$

$$\frac{40 \ psi}{288 \ K} = \frac{p_2}{303 \ K}$$

$$p_2 = 42.1 \ psi$$

This may not seem like a significant difference, but when a few tenths of a psi mean a major change in the ride height, the size of the contact patch (the part of the tire in contact with the road surface), and the effective suspension spring rate, this change can be significant. Furthermore, the temperature of a typical racing tire at use on an asphalt track is 200°F–220°F, or roughly 373 K. This increase in temperature further alters the tire pressures, building our example up to a pressure of 51.8 psia. Clearly, this type of behavior must be accounted for before the tires go on the car, which means the predictive power of Gay-Lussac's law is very important in racing. The exact same effect can be observed by monitoring the tire pressures throughout the year on your passenger car. If you fill the tires to 35 psig on a warm day in the summer (90°F or 305 K), you will find that the pressure has dropped significantly on a cold winter day. Assuming the winter afternoon high is 20°F in January, Gay-Lussac's law says that the pressure will have dropped by 4–5 psi, or nearly 13%!

$$\frac{p_1}{T_1} = \frac{p_2}{T_2}$$

$$\frac{p_1}{305 \ K} = \frac{p_2}{266.4 \ K}$$

$$p_2 = 0.87 \ p_1$$

1.3   Struts, shock absorbers, and Boyle's and Charles's laws

Chemistry Concepts: gas laws, kinetic molecular theory, first law of thermodynamics

Expected Learning Outcomes:

- Use a simple piston and kinetic molecular theory to describe simple gas laws

- Understand that for a gas at constant temperature, its pressure and volume are inversely proportional
- Understand that for a gas at constant pressure, its volume and absolute temperature are proportional
- Explain how a shock absorber dissipates energy using gas laws, the concepts of heat and work, and kinetic molecular theory

Pistons play critical roles in many automotive systems, from the engine itself to the brakes and suspension. Pistons are also essential to demonstrate critical relationships between other gas properties, namely the links between pressure and volume and temperature and volume. In this section, we introduce Boyle's and Charles's laws using concepts from kinetic molecular theory and a simple piston. Then we use these concepts to explain how a typical gas-filled shock absorber or strut functions in your car's suspension, and why you use a liquid in your brake lines rather than a gas.

The simple piston is an idealized system where a fixed quantity of gas is trapped within a chamber sealed on one side by a piston in contact with the atmosphere. The chamber functions as a closed system, and thus fixes the number of molecules of gas within the system, defined here as gas trapped within the chamber. The remainder of the apparatus lets temperature, pressure, and volume vary. If we hold one of these variables constant (temperature, pressure, or volume), the relationship between the other two variables can be explored via experiment. The key element to each experiment is understanding that the piston itself will not move when the pressure inside the chamber is equivalent to the pressure outside the chamber. This criterion of equal pressures establishes the equilibrium position of the piston.

If one holds the temperature constant, the pressure in the system can be varied by changing the external force applied to the piston (Figure 1.2). Since the piston surface area in contact with the system is invariant, then applying increased force to the outside surface of the piston causes the external pressure to rise. The system responds by reducing in volume until the pressure inside the chamber is equal to the pressure outside the chamber. You can observe the inverse phenomenon at home using a syringe or a turkey baster. Pull back the plunger, plug the hole with your finger, and push on the piston or bulb. Do you feel the extra pressure as the gas volume is compressed? Kinetic molecular theory tells us that the origin of the increased pressure is an increase in the number of collisions between the gas molecules and the chamber boundaries in a given increment of time. Fixing the temperature fixed the average kinetic energy in the system and thus the average energy of a collision. Since we can't give the particles more kinetic energy via heating to combat the increased external pressure, the gas is forced to increase the number of collisions, and does so by decreasing in volume. Compressing the gas leaves less free space for the gas particles, and if there is less free space, the number of particle–particle

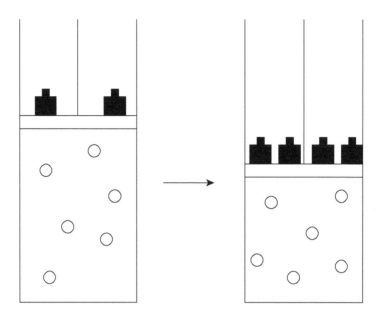

Figure 1.2 Simple piston illustrating Boyle's law. The system pressure must balance the external pressure to keep the piston stationary. As the external force and pressure increase, the system compensates by decreasing in volume at constant temperature and constant composition.

and particle–wall collisions must increase. The fact that volume decreases as pressure increases indicates that these properties must have an inverse relationship, and that relationship is indicated in Boyle's law.

$$p \propto \frac{1}{V}$$

$$p_1V_1 = p_2V_2$$

Instead of fixing the temperature, if we hold the external pressure constant, we can explore the relationship between volume and temperature using our simple piston (Figure 1.3). If we heat the gas trapped inside the chamber, the average kinetic energy of the particles and the average energy of a collision increase according to kinetic molecular theory. Not only is each collision more energetic, but the greater kinetic energy leads to a greater number of collisions between the gas particles inside the system and the inside surface of the piston in a given time interval. Both effects cause a greater force to be exerted on the piston by the gas particles trapped within the heated system. Since the external force remains constant, the piston must move up to find a new balance point.

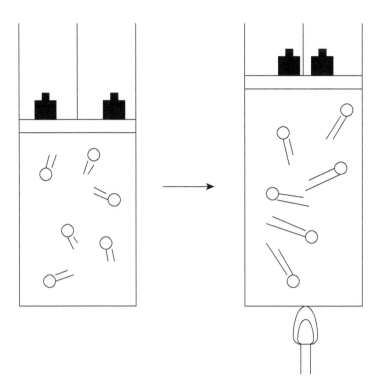

Figure 1.3 Simple piston illustrating Charles's law. At a constant composition and external pressure, heating the system causes the gas to expand as the gas molecules take on extra kinetic energy and produce more energetic collisions with the walls.

Effectively, it moves to a height at which the collision frequency decreases to the point that the internal and external forces on the piston are balanced again. Upward motion of the piston leads to an increase in the gas volume, which tells us that gas volumes increase with increasing temperature. This relationship is known as Charles's law.

$$V \propto T$$

$$\frac{V_1}{T_1} = \frac{V_2}{T_2}$$

Together, Boyle's and Charles's laws tell us a lot about the way a gas-filled shock absorber works on a car (Figure 1.4). When your car hits a bump in the road, the force of that collision is transferred almost entirely to the car. Rarely do you observe the road move or give way as a result of these collisions. If the car cannot absorb and dissipate that energy in some way,

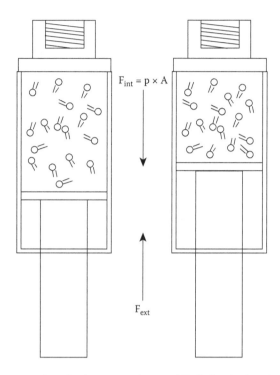

$$F_{int} = p \times A$$

$$F_{ext}$$

Figure 1.4 Cross-sectional schematic of a gas-filled shock absorber and its relationship to the gas laws. The external force when the car is parked (like the situation on the left) is simply the weight of the vehicle, which is balanced by the pressure in the gas chamber. When the car wheel impacts something on the road surface, the external force becomes the weight of the vehicle plus the force of the impact. The additional force is offset by a reduction in the chamber volume, which results in an increase in the pressure to offset the impact force.

it will be transferred to the passengers in the vehicle, who will have a very uncomfortable ride. Softening the impact is one of the main roles that the springs and shock absorbers in your suspension play, and shock absorbers do their part by dissipating the impact energy following the gas laws and principles of kinetic molecular theory that we have discussed to this point. In a gas-filled shock absorber, the force of the impact transferred to the shock absorber is applied to a piston in contact with a fixed number of gas particles confined in a chamber, much like the simple piston model. Since the shock absorber is in contact with a high flow of atmospheric gas and constructed almost exclusively of materials with high thermal conductivity like metals, it is reasonable to assume that the temperature of the shock absorber and the gases it contains will remain relatively constant. The force of impact on the external side of the piston will increase the external pressure, and as we now know, the system will respond by decreasing in volume so that the impact force is opposed by the increased number of collisions between confined gas particles and the walls of the shock absorber. Effectively, some of that impact energy is stored within the compressed gas and is reflected as a change in the gas pressure and temperature. The high pressures cause the gas to flow through small openings within the shock absorber that are activated at high pressure and expand, which results in cooling of the gas, equilibration of gas pressure within the shock absorber, and dissipation of the energy the gas absorbed. Of course, it is reasonable to assume that some of the energy stored as heat will be dissipated as cool air rushes over the exterior of the shock absorber. After the impact event and expansive cooling within the shock absorber, the spring will force the shock absorber to return to its equilibrium position, and the absorber design allows the gas to refill the chamber without a significant energy cost. The combined effects of the gas behavior—made possible with special combinations of one-way and two-way valves inside the shock absorber— is that much of the energy is dissipated, allowing the passengers less of a jolt and a more comfortable ride.

Magnetorheological Fluid Shock Absorbers

Many automobile manufacturers have recently begun high-lighting "magnetorheological fluid" shock absorbers in television advertisements for performance and luxury cars. Patents are also filed for magnetorheological fluid-based engine dampers and continuously variable transmissions. But what are these magnetorheological fluids, and how do they differ from the gas-filled designs discussed in this section of this text? In magnetorheological fluids, magnetic particles are suspended in a carrier fluid (typically an oil). When a magnetic field is applied to this oil, the magnetic particles become organized and oriented to some degree, changing the viscosity of the fluid. In other words, thanks to the suspended magnetic particles, electromagnets can be used to alter the resistance of the fluid to flow. In a car, this means that the stiffness of a suspension and the rate that a shock dissipates energy can be controlled by an electromagnet or series of electromagnets surrounding the fluid reservoir in the shock absorber. The magnetic particles are typically ferromagnetic iron or iron alloys with particle sizes on the order of 10 microns or less. These particles are coated with molecules that cause the Fe particles to be soluble in the carrier fluid. These molecules will typically have a functional group that can coordinate to the Fe atoms on the surface of the particles—and a long hydrocarbon tail. Such molecules are called surfactants and are discussed in greater detail in Chapter 4. The surfactant-coated particles are then suspended in the carrier oil and are ready to use. Patents report magnetorheological fluids that can change viscosity by up to four orders of magnitude, depending on the strength of the applied magnetic field, which clearly gives drivers an enormous degree of control over the feel of their vehicle. Because of the degree of control over the suspension performance, it is likely that magnetorheological suspensions will become more common in the coming years.

Pistons also are used in cars to squeeze the brake calipers, forcing the friction material in contact with the brake rotors or drum surface and dissipating the vehicle's kinetic energy via friction and heat. However, brake lines in cars are filled with hydraulic fluid rather than a gas, and kinetic molecular theory and the gas laws tell us why. Gases are the most compressible state of matter, as indicated in point one of KMT and in Boyle's and Charles's laws. If you had a gas in your braking system, stepping on the brake pedal would simply compress the gas without actuating the piston in the brake calipers. You would need a significant gas compression to generate a large enough increase in the number of collisions to generate the squeezing force required to stop the vehicle. On the other hand, liquids, which are far less compressible, do not experience a significant change in volume when you step on the brake pedal. As a result, the force of your pedal push is transferred more effectively to the brake calipers and is plenty strong enough to actuate the caliper piston.

1.4 Ideal gases

Chemistry Concepts: perfect gas, ideal-gas law, vapor pressure, partial pressure

Expected Learning Outcomes:
- Describe the relationships between perfect gases and kinetic molecular theory
- Write the ideal-gas law and explain the relationships between variables
- Suggest ways to alter the performance of a suspension based on the gas laws

Rather than invoking the simple relationships between gas properties when confronting a new problem involving gases in cars or in chemistry, it is beneficial to discuss the physical properties of gases using a combined model known as the ideal-gas law. Ideal (or perfect) gases are those that are well described by kinetic molecular theory. For example, ideal gases are assumed to exhibit no attractive or repulsive interactions between gas particles and to occupy a negligible volume within a container, essentially identical assumptions to those that were made when developing kinetic molecular theory in Section 1.1. Although these assumptions seem quite unrealistic, they describe nearly perfectly the behavior of monatomic noble gases and deviate only slightly from the true behavior in many gas systems. A detailed discussion of the ideal-gas law and various manipulations of it are typically provided in general chemistry textbooks, and thus our discussion here is quite brief.

The ideal-gas law can be written as follows, and one can easily show that it reduces to the simple gas laws discussed previously if we make similar sets of assumptions:

$$pV = nRT$$

Here, the variables are defined as they have been earlier in this chapter, with the addition of n to represent the number of gas molecules in the system using the mole concept (1 mole of anything is equivalent to 6.02×10^{23} entities) and R as the ideal-gas constant, which is related to the Boltzmann constant and Avogadro's number. One can determine the value of R experimentally by measuring the volume of 1 mole of a gas at standard temperature and pressure. To generate Boyle's law, we simply consider the properties of a gas before and after a change in pressure. In a closed system that is held at constant temperature (the same assumptions in our discussion of the proof for Boyle's law), the entire right-hand side of this equation is identical no matter what the specific temperature and pressure, since all terms are constant. Thus, the product of the initial pressure and volume must be equal to the product of the final pressure and volume.

$$p_1V_1 = nRT$$
$$p_2V_2 = nRT$$
$$\therefore p_1V_1 = p_2V_2 = nRT$$

Similar proof exercises can be performed for Charles's and Gay-Lussac's laws. The beauty of the ideal-gas equation is that we can consider more complicated situations than simply binomial relationships between variables. We can also generate all of the binomial relationships from this combined gas model.

Let's use the ideal-gas law to design a "stiffer" suspension that returns the contact patch on our tires to the ground more quickly after an impact and resists variations in the ride height as the weight distribution is changed during cornering or heavy braking. Really, what we desire is a shock absorber that experiences smaller changes in volume for a given increase in the external force, which intuitively means that we need a higher equilibrium gas pressure in our shock absorber. Thus, we should rearrange the ideal-gas equation so that we see what variables influence the pressure in the piston:

$$p = \frac{nRT}{V}$$

A typical car suspension is designed with very little control over the temperature of the shock absorber. Thus, let us consider the temperature to be constant for the remainder of this thought exercise. We thus have two ways to increase the equilibrium pressure in our shock absorber: increase the number of gas molecules in the chamber when it is constructed or change the chamber volume. Certainly both are possible, but most commercial adjustable shock absorbers rely on the pressure/volume relationship. When you adjust an adjustable shock absorber, you are really changing the equilibrium height of the gas chamber. Since the cross-sectional area of most shock absorbers is fixed (though you could certainly design a shock absorber with a different cross-section area to obtain the same effect) and volume is the product of area and height, adjusting the height results in a volume change in the piston and an increased equilibrium pressure in the shock absorber, providing a "stiffer" suspension.

Nitrogen-Filled Tires

Another important gas property is Dalton's law of partial pressures, which says that the total pressure in a gas system is equivalent to the sum of the pressures due to the individual gas components. For example, the atmosphere is roughly 21% oxygen gas and 79% nitrogen gas. At the surface of the Arctic Ocean, the total atmospheric pressure is roughly 1 atm, with a pressure due to oxygen gas of 0.21 atm and a pressure due to nitrogen gas of 0.79 atm. The pressures due specifically to oxygen and nitrogen are called partial pressures. Tires are typically filled with air, meaning that the air in your tire is also roughly 21% oxygen and 79% nitrogen. When tire installers offer to fill a car's tires with 100% nitrogen gas, effectively they are altering the partial pressures in your tire to substantially reduce the partial pressure of oxygen gas and increase the partial pressure of nitrogen. Doing so provides two arguably minor advantages, at least with respect to domestic driving applications. First, by reducing the oxygen partial pressure, you reduce the chance of oxidation reactions (see Chapter 3 for more on oxidation) occurring inside the tire that contribute to stiffening and decay of the rubber. The second advantage relates to a reduction in the number of water molecules in the filled tire. Air always contains a nonzero partial pressure of water, and the mechanical compressors used to fill tires tend to generate very wet air unless special steps are taken to dry the compressed gas. If you own a compressor, turn it on in the summertime and place your hand over the outlet to experience this wet air firsthand. The liquid water that enters the tire as a result of the compressor will partially vaporize, and the extent of vaporization is related to the tire and wheel temperatures. Since these values vary substantially during driving, the presence of liquid water causes more significant pressure changes in your tires as they heat and cool. The presence of liquid water in the tires also contributes toward chemical degradation of the wheels and tires. By filling them with dry nitrogen, you prevent liquid and gaseous water from entering the tire, combating both of the water-related drawbacks.

1.5   Air bags

Chemistry Concepts: gas laws, reaction stoichiometry, kinetics, thermodynamics

Expected Learning Outcomes:
- Explain the roles of the various chemical components in an airbag deployment system
- List criteria to which an effective air-bag system must adhere
- Describe the chemical mechanism behind deployment of a first-generation air bag, with a particular focus on reaction rates

Another place where basic gas behavior is essential to the function of modern automobiles is in the supplemental restraint system, typically called the *air-bag system*. The major purpose of an air bag is to prevent an extremely serious injury by cushioning one's head and/or body in an impact. An air bag must deploy and fully inflate very quickly (less than 0.04 s), which is so fast that the only reasonable inflation mechanism is a very rapid gas-producing chemical reaction. Using a chemical reaction–based deployment system also introduces other criteria that must be accounted for in terms of safety and environmental responsibility. For example, the reaction cannot start until the reaction is needed, otherwise random deployment of air bags would occur, causing needless injuries. The overall deployment process must dissipate heat very quickly. To be fast enough, the reaction is likely to be very exothermic, but the air bag cannot get so hot that it burns the passengers (we will discuss heat transfer in Chapter 5). The chemicals involved in the reaction must be very stable; they cannot oxidize in the atmosphere or absorb atmospheric moisture, both of which would prevent effective initiation of the reaction in an emergency. The chemicals involved and products they produce should also be nontoxic. The original air-bag chemicals do not meet this final requirement, but newer generation air-bag systems have redesigned chemistry that limit exposure of people or the environment to toxic chemicals and toxic by-products.

The original air-bag deployment chemistry is a three-reaction process involving sodium azide (NaN_3), potassium nitrate (KNO_3), and silica (SiO_2). When an impact occurs, it triggers an electric match or fuse that provides the energy necessary to initiate decomposition of the sodium azide (Figure 1.5). This is the step that generates a majority of the gas to inflate the air bag. As you can see from the reaction stoichiometry following this paragraph, 3 moles of nitrogen gas are generated for every 2 moles of sodium azide that are decomposed. Unfortunately, this reaction also generates sodium metal as a product, which is highly reactive with moisture and potentially dangerous. The second reaction in the process generates some additional nitrogen gas, but its primary role is to protect the passengers from the sodium metal by converting the sodium to a more stable but still fairly reactive oxide. The final step in the sequence

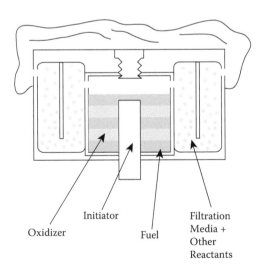

Figure 1.5 Schematic of a typical air-bag inflation reaction chamber. In the traditional air bag, the fuel is the sodium azide and the oxidizer is the potassium nitrate. (Adapted from T. Mooney, G. B. Little, and G. Little, "Inflator for Vehicular Airbags," US Patent 5,713,595, filed Oct. 31, 1994, and issued Feb. 3, 1998.)

uses SiO_2 to convert the relatively reactive potassium and sodium oxides into relatively inert alkali silicate glasses. These very finely divided glass particles are not capable of causing acute injuries to the passengers and undergo no dangerous chemistry in the environment.

$$2NaN_3 \rightarrow 2Na + 3N_2(g)$$

$$10Na + 2KNO_3 \rightarrow K_2O + 5Na_2O + N_2(g)$$

$$K_2O + Na_2O + 2SiO_2 \rightarrow K_2SiO_3 + Na_2SiO_3$$

The patent literature shows that a typical driver's air bag is a nylon fabric bag that has a volume of 35 to 70 L, which means our gas-generating reactions must produce at least this volume of gas for the air bag to fully deploy. We can check whether an appropriate amount of nitrogen gas is produced in a typical air bag by using the gas laws that we have discussed in this chapter and by making some simple though not entirely accurate assumptions about the air-bag system. Studies have shown that the peak pressure in a deploying air bag after it emerges from the steering column is roughly 35 kPa above atmospheric pressure, or approximately 1.35 atm. It is also known that the electronic match that initiates the azide decomposition does so by producing a temperature of 300°C. If we assume that the nitrogen behaves like an ideal gas, that our air bag contains 65 g of sodium azide, and that the gases generated initially are at the initiation temperature, then the gas laws tell us that the ≈1.5 moles of nitrogen gas generated in the first step of the deployment reaction sequence will expand to a volume of 52.3 L.

$$65 \text{ } g \text{ } NaN_3 \times \frac{1 \text{ } mol \text{ } NaN_3}{65.019 \text{ } g \text{ } NaN_3} \times \frac{3 \text{ } mol \text{ } N_2}{2 \text{ } mol \text{ } NaN_3} = 1.50 \text{ } mol \text{ } N_2$$

$$V = \frac{nRT}{p}$$

$$1.50 \text{ } mol \text{ } N_2 \times \frac{0.0821 \text{ } L \text{ } atm \text{ } mol^{-1} \text{ } K^{-1} \times 573 \text{ } K}{1.35 \text{ } atm} = 52.3 \text{ } L \text{ } N_2$$

This does not seem to be enough gas to fully inflate many driver-side air bags given the previously stated volume range. However, note that this exercise was performed considering only the nitrogen generated by the azide decomposition step and discounting the nitrogen produced in the second step of the reaction sequence. Furthermore, all three of the reactions involved in the traditional air-bag deployment chemistry are exothermic, meaning that they release energy in the form of heat. Therefore, the initial temperature of the nitrogen gas generated in this process will be much higher than 573 K. Since temperature is in the numerator of our expression, an increase in temperature will lead to an increase in the volume of nitrogen gas produced, suggesting that this amount of sodium azide will be appropriate to fill the air bag.

Since so much heat is generated during the air-bag deployment reactions, it is reasonable to wonder if the gases are still hot when the passenger makes contact with the air-bag surface. We can also use the gas laws to answer this question. The gases generated during deployment must escape from the air bag so that it does not stay inflated and risk trapping or suffocating the passengers, and the gases do so by emerging through small holes in the air-bag surface. From the previous case study, we know that the pressure drop across the air-bag surface is 35 kPa at its maximum. Since the volume of the air bag remains a constant 65 L, we can use Gay Lussac's law to determine how much cooling occurs due to the expansion of the gas as it exits the holes in the air bag. This analysis also requires that we keep the number of gas molecules in the air bag constant, which is valid during the initial stages of deployment when more gas is generated than is required to inflate the bag. Assuming the atmospheric pressure is 1.0 atm, then:

$$\frac{p_1}{T_1} = \frac{p_2}{T_2}$$

$$\frac{1.35\ atm}{T_1} = \frac{1.00\ atm}{T_2}$$

$$T_2 = \frac{1.00}{1.35} T_1 = 0.74 T_1$$

Thus, as the gases exit the air bag, the temperature drops by roughly 25%. The gases should also experience additional expansion-related cooling as the air bag breaks free from its storage chamber. Research has shown that there can be a pressure drop of over 400 kPa in the instant after the bag breaks free (remember that the bag volume is much less than 65 L when compressed within the air-bag chamber), and a similar analysis with Gay Lussac's law shows this is associated with a roughly 75%–80% drop in the gas temperature immediately following the reaction sequence. Clearly, this expansion-related cooling is suitable for protecting the passengers from the high temperatures reached within the reaction chamber.

As mentioned earlier, a successful air-bag system must not involve toxic chemicals, and the sodium azide used in the original air-bag designs is a poison that targets the central nervous system and respiratory system. It can be absorbed through the skin, lungs, and ingestion, and is known to generate toxic gas by-products. For this reason, many alternative systems for generating gas in vehicular air bags have been proposed. Alternative systems include the use of less toxic nitrogen-bearing organic materials such as imidazoles and hydrazides to generate nitrogen gas as well as systems generating predominantly CO_2. Providing a full list of alternatives is beyond the scope of this text, but one can learn about the details of various deployment chemistries by viewing the 1972 patent by Schneiter[*] and subsequent patents citing this work.

---

[*] F. E. Schneiter, "Gas Generator," US Patent 3,692,495, filed June 19, 1970, and issued Sep. 19, 1972.

chapter two

Combustion, energy, and the IC engine

The core of any modern automobile is the power train (Figure 2.1), which is a series of devices that convert chemical potential energy to kinetic energy and do the work of moving the vehicle. The power train is made up of the engine, transmission, differentials, driveshafts/half shafts, the axles, and the wheels. The engine is the device that converts chemical potential energy into kinetic energy via rotation of the crankshaft and flywheel. The transmission transmits the kinetic energy of the flywheel to the driveshaft or half shafts and exerts the greatest control over the final drive ratio, or the number of flywheel rotations per rotation of the wheels. Essentially, it facilitates smoother and quicker acceleration. In some vehicles, power is transferred from the transmission directly to half shafts and then to the wheels via the constant-velocity joints, which are a special bit of engineering that allows uninterrupted power transfer to the wheels no matter the positional state of the suspension. In other vehicles, the transmission rotates a driveshaft that sends the power to a differential gear assembly. The differential converts the kinetic energy of driveshaft rotation into rotation of the axle(s) oriented 90° with respect to the driveshaft. The rotating axle or axle half shafts then cause rotation of the wheels. We will visit chemical concepts related to the transmission and various other drivetrain components in subsequent chapters. However, in this chapter, we focus on the engine itself, the reactions and fuels it uses to generate power, and the ways that we can adjust the power output of an engine by varying the engine chemistry and combustion stoichiometry.

2.1 Fossil fuel–based engines

Chemistry Concepts: combustion, types of energy, properties of gases, gas laws, thermodynamics, properties of metals

Expected Learning Outcomes:

- Understand the purpose of an engine—to harness chemical potential energy and use it to do work
- Explain the difference between internal combustion gasoline and diesel engines
- Explain the typical internal combustion engine four-stroke cycle
- Communicate the general definition of a combustion reaction

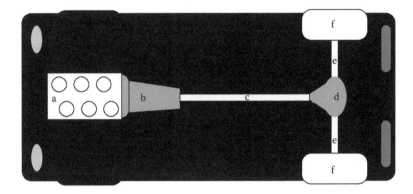

Figure 2.1 A rear-wheel-drive automobile power train. The power train includes the engine (a), transmission (b), driveshaft (c), differential (d), half shafts (e), and wheels/tires (f).

There are two main types of fossil fuel engines found in commercial vehicles: the internal combustion (IC) diesel engine and the internal combustion gasoline engine. They both harness chemical potential energy by performing oxidation reactions of organic-based fossil fuels, though diesel engines can be modified to run on more-renewable organics such as vegetable oil or biodiesel. There is also another type of fossil fuel engine called a rotary engine that is used far less frequently (at the writing of this book, Mazda is the only major automobile company putting Wankel rotary engines in consumer vehicles), and automobile manufacturers are pursuing direct-injection gasoline engines. Rotary engines operate on essentially the same chemical principles as internal combustion engines, but use a different mechanical mechanism for navigating through the combustion cycle. Direct-injection gasoline engines are the topic of a sidebar later in this chapter, but are essentially diesel-like engines that run on gasoline. Natural gas engines have also begun to appear in commercial vehicles, primarily in mass transit applications such as bus systems. However, prototype natural gas passenger vehicles have been developed by Honda, Toyota, Audi, BMW, Chevrolet, Citroen, Fiat, Ford, Mercedes-Benz, Opel, Peugeot, Renault, Volkswagen, and Volvo as of 2013. There are also non-fossil fuel–based engines, such as the electric motors found in fuel–cell vehicles and full-to-partial-electric vehicles. However, in these types of power plants, chemistry is still crucial in explaining their operation. The power sources are based upon either light interactions with matter or chemical potential energy, albeit in a different type of reaction than in a fossil fuel engine. Hybrid-electric vehicles and in most cases electric vehicles should be considered fossil fuel–based, though this energy is often stored in the chemistry of a battery. Hybrids typically charge their batteries using a small fossil fuel–based engine that also assists with hard acceleration and high-speed operation. Plug-in electric vehicles are often charged via the commercial power grid, which is currently dominated by coal combustion and other fossil fuel–based electrical plants. Electric vehicles charged by wind, nuclear, and other power sources or by solar charging stations are not fossil fuel–based. This chapter focuses on the two most common types of fossil fuel–based commercial engines in automobiles, with a greater emphasis on the gasoline internal combustion engine.

The modern gasoline internal combustion engine is a four-stroke engine that harvests the energy released by a combustion reaction and converts it to kinetic energy. While engines and the associated accessories contain many parts, at this point we will focus only on those parts involved in the combustion process itself (Figure 2.2). An engine must initiate and contain a combustion reaction, and it accomplishes both of these tasks in a reactor vessel that is called a *combustion chamber*. The combustion chamber is cylindrical in shape when the piston is at the bottom of a downstroke and has a geometry controlled by the shape of the cylinder head (hemisphere, wedge, etc.) when the piston is near the top of its stroke. Each cylinder of the engine corresponds to one combustion chamber, and typical IC engines have between 3 and 12 cylinders. The rounded walls of the combustion chamber are composed of metal liners that slide into cylindrical holes cut or cast in a block of metal. The block is typically iron or aluminum, and the liners are nearly always iron or steel. The bottom

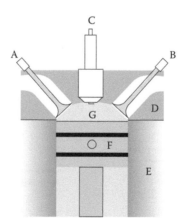

Figure 2.2 Schematic of a typical combustion chamber. There are intake (A) and exhaust valves (B) in the cylinder head (D), as well as a port for the spark plug (C). The head is mated to the block (E) via a thin gasket, usually made of copper or another metal. The bottom of the chamber is the top of the piston (F). The combustion chamber (G) is shown here near the end of an upstroke at its smallest possible volume.

of the combustion chamber is the top of a piston that can move perpendicular to the cylinder radius. The bottom of the piston is connected to a shaft called the crankshaft by connecting rods and pins. The top of the combustion chamber is part of the cylinder head, which contains plumbing for the air and fuel intakes as well as the exhaust, valve ports for the valves that control the flow of fuel and combustion gases in and out of the combustion chamber, a port for the spark plug that initiates the combustion process, and plumbing for lubricating oil and engine coolant (typically). The valves are opened and closed at specific, controlled times by either pushrods, rocker arms, and valve springs actuated by an internal camshaft or directly by the camshafts mounted over the tops of the valves in the cylinder head. Camshafts rotate with the engine and contain teardrop-shaped lobes that can push the valves open smoothly for a short period of time.

A combustion cycle begins by the downward stroke of the piston, enlarging the volume of the combustion chamber and creating a pressure drop in the cylinder (Figure 2.3). The intake valve opens during this stroke, which connects the combustion chamber with the atmosphere. Atmospheric gases are drawn into the cylinder until the pressure within the cylinder is equal to the atmospheric pressure in the air intake pipe (remember our volume/pressure relationship from Chapter 1?). Fuel is also introduced into the chamber during this stroke, typically via a device called a fuel injector that generates an aerosol of gasoline in the inlet air, which is a fine mist of rapidly evaporating tiny fuel droplets. The end of this cycle is when the intake valve closes. The second stroke of the piston is an upward motion that compresses and heats the fuel–air mixture by reducing the volume of the combustion chamber. Just before the end of

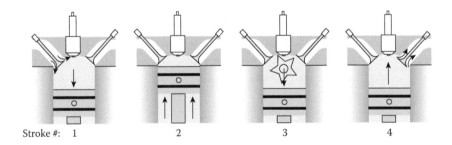

Stroke #:   1   2   3   4

Figure 2.3 Schematic of behavior in the cylinder during the four-stroke cycle. From left to right, downward stroke #1 brings air and fuel into the combustion chamber. Upward stroke #2 compresses the fuel–air mixture. The firing of the spark plug initiates combustion, which forces the piston downward in stroke #3. The exhaust valve opens, and upstroke #4 clears the cylinder of combustion products and uncombusted fuel.

the second stroke, the spark plug fires, providing the energy to initiate combustion of the fuel. The combustion reaction generates many more moles of gas than were in the cylinder after the first two strokes, and the chemical energy released as heat causes the gases to rapidly expand. The expanding gases push the piston down in its third stroke of the four-stroke cycle, and the downward motion of this stroke is what actually provides energy to turn the crankshaft and power the vehicle. The fourth stroke of the cycle is upward motion of the piston, which occurs with the exhaust valve(s) open. Here, the upward motion of the piston forces the combustion products and unreacted gases out the exhaust valve and into the exhaust system. This concludes the four-stroke cycle and leaves the piston in the position to begin a new four-stroke cycle. During these events, the crankshaft rotates two full turns, and chemical potential energy is being captured during only one-half turn of the crankshaft for each cylinder (the downward stroke #3 after combustion). The rotating crankshaft is attached to a large metal disk called a flywheel. The flywheel is where the engine and transmission meet and where begins the process of transferring the kinetic energy of the rotating flywheel to kinetic energy of the wheels that drive the vehicle.

Diesel engines work in a similar four-stroke fashion, but differ significantly in the location where fuel and air are mixed, the method used to initiate the combustion reaction, and the chemical makeup of the fuel (see Section 2.2). The basic layout of a diesel engine is the same as the IC gasoline engine, but the first stroke draws in only air rather than a mixture of air and fuel. The compression stroke (stroke #2) in both engines leads to significant heating of the gases in the chamber; however, a diesel engine has much higher compression ratios, meaning that the gases are compressed to a greater degree. In fact, the diesel engine compresses the air so much that its temperature is high enough to induce combustion of the fuel without an external ignition source. The fuel is sprayed into the hot compressed air just before the end of stroke #2 and during part of downstroke #3, combusting fully and completely in the uniformly hot air. Diesel engines also use much lower fuel/air ratios, ensuring that oxygen is present in excess and that the fuel is combusted more completely, since it is the limiting reagent. This, and the fact that fuel mixes with air in the cylinders of a diesel, giving diesel engines the ability to inject fuel over a greater portion of the downward third stroke, help to make diesels efficient and powerful. The final upstroke accomplishes the same tasks as the gasoline internal combustion engine, namely driving out the exhaust gases and preparing the cylinder for another cycle. If a diesel engine is to be used in an environment where the outside air is very cold, glow plugs are used to raise the initial temperature of the air and aid in the startup cycle of a diesel engine, which requires a minimum air temperature to operate properly.

Direct-Injection Gasoline Engines

Diesel engines offer advantages in terms of efficiency
(Section 2.4) and torque (Section 2.5), but diesel combustion generates additional particulate pollutants and offers
more challenges for a vehicle's catalytic converter system
(Section 3.3). Diesel fuel is also more expensive than gasoline
on a per gallon basis, since it takes more energy to recover and
refine diesel from crude oil (Section 2.2). But what if you could
get the same benefits of diesel, e.g., extremely high gas mile-
age and large torque outputs, in an engine that burns regular
gasoline? Hyundai Motor Company, Mitsubishi, Audi, Mazda,
General Motors, and others have paired with suppliers like
Delphi Electronics and Bosch to develop direct-injection four-
stroke gasoline engines that permit the combustion of gasoline without the need for spark plugs. Historically, it has been
difficult to build engines that can withstand the required
pressures and temperatures for direct-injection gasoline combustion, and the required fine-tuning of fuel/air ratios, fuel
delivery, etc., was not possible before the advent of complete
engine management systems with computer control. Thanks
to the availability of very fast microprocessors, engineers have
recently developed management programs that can successfully operate the direct-injection gasoline engine. Likewise,
modern materials have led to fuel injectors and other components that can withstand the very high temperatures and pressures required. Many of the direct-injection gasoline engines
also take advantage of turbo or superchargers (or both!) in
combination with very high compression ratios to generate
cylinder air temperatures hot enough to initiate gasoline
combustion without a spark. Other than these engineering
differences, the chemistry of direct-injection gasoline engines
is identical to that of internal combustion gasoline engines.
Truthfully, the only difference with chemical consequences
is the very high fuel/air ratio in the direct-injection engines,
which helps to ensure that the fuel is the limiting reagent,
leading to high combustion efficiency. The U.S. government
estimates that direct-injection gasoline engines may offer
efficiency improvements over conventional gasoline internal
combustion power plants of up to 12%, the single largest-
impact technological advance for fossil fuel–based engines.
Expect to see additional direct-injection gasoline engines on
the market in the coming years.

Many of the engine behaviors are well describable using the gas-law concepts discussed in the previous chapter. For example, knowing that gases fill their containers uniformly and the pressure/volume relationship from Chapter 1 explains precisely why air rushes into the cylinders during stroke #1: Increasing the combustion chamber volume drops the pressure according to Boyle's law, and the open valve allows the cylinder to contact the atmosphere. The gas-phase fuel–air mixture has no choice but to rush in and equilibrate the pressure between the air intake and the combustion chamber. Compression during stroke #2 leads to gas heating and an increase in the pressure of the chamber. This step is a little more difficult to understand as pressure, temperature, and volume are all varying simultaneously, and as such our simple gas laws (where two variables from P, V, T, and n must be held constant) break down. However, if the volume decrease during compression does not fully offset the pressure gain, the ideal-gas law does show us that the temperature must increase to compensate for this offset. Expansion of hot gases as described in Charles's law is the precise mechanism by which a portion of the energy released by combustion is captured by the engine and converted to the mechanical kinetic energy that is used to drive the vehicle (i.e., the driving force behind the piston downstroke #3).

The next section examines in detail the chemical reactions involved in these combustion processes.

2.2   The combustion reaction

Chemistry Concepts: stoichiometry, thermochemistry, chemical reaction basic definitions, gas laws
Expected Learning Outcomes:
- Perform thermochemical calculations using combustion reactions to understand the energy generated in an IC engine
- Describe the properties of a good fuel
- Write a generic combustion reaction and list the common types of emission by-products

By definition, a combustion reaction is any reaction between a material and an oxidant [typically O_2 (g)] that releases energy in the form of heat (Section 2.4). Chemical reactions are processes that alter the bonding types and relative positions of atoms within molecules. Starting materials are called *reactants*, and the final materials following the rearrangement are called *products*. In a combustion reaction, the nonoxidant starting material is called a *fuel* and can be a variety of chemical compounds. Most commonly, combustion is introduced in general and organic chemistry as the reaction of hydrocarbon fuels with oxygen gas to produce carbon dioxide and H_2O. However, the typical organic-based fuel contains more

Table 2.1 Representative Reactions of Gasoline Combustion
Using Simple Molecules

Molecule types	Reactions	Enthalpy release $(-\Delta_c H)$
Hydrocarbons	$CH_4 + 2O_2 \rightarrow CO_2 + 2H_2O$	890.8 kJ/mol
Nitros, amines	$CH_3NO_2 + {}^5/_2O_2 \rightarrow CO_2 + {}^3/_2H_2O + NO$	747.6 kJ/mol
Sulfur-bearing organics	$CH_3SH + 3O_2 \rightarrow CO_2 + 2H_2O + SO_2$	1239.2 kJ/mol
Oxygenated organics	$CH_3OH + {}^3/_2O_2 \rightarrow CO_2 + 2H_2O$	763.7 kJ/mol

elements than just carbon and hydrogen and produces other gases besides simply carbon dioxide and water. Table 2.1 contains a series of simple combustion reactions involving elements commonly found in organic fuels that highlight the types of reactions and combustion products that are formed in an internal combustion (IC) engine. Combustion engines also produce unburned hydrocarbon fuels and what is called thermal NO_x, gases with the formula of NO_x that form when atmospheric nitrogen becomes very hot and reacts with atmospheric oxygen. We will further discuss exhaust emissions and how they are reduced in IC vehicles in Section 3.3.

A majority of internal combustion gasoline engines and diesel engines are designed to use specific fractions of hydrocarbons distilled from crude petroleum as their fuel. Petroleum, or crude oil, is essentially a complex mixture of organic chemicals left over from the degradation of buried marine microorganisms. Only the lighter, more volatile components of petroleum have the proper characteristics to be a fuel source for cars, and these components are separated from the crude oil using a common chemical-separation technique known as distillation. In laboratory distillation, a liquid is heated to boiling, and the hot vapors are passed through a cooling device called a *condenser* (Figure 2.4), where they cool, condense, and are collected in some type of container. In industrial distillation, the hot vapors are allowed to contact the cooler liquids in a large device called a *distillation tower*, which is a large metal cylinder often filled with plates that increase the liquid–vapor contact area to improve mass transfer between phases. In a distillation tower, different fractions of the crude oil condense at different heights, depending on their boiling point (and several other factors to a lesser extent). Light, volatile compounds remain gases throughout the column and exit as vapor near the top, where they can be collected or passed back through the tower as liquids after passing through a condenser. Heavier materials with higher boiling points condense further down in the column or tower, and may be removed as liquids at any height when plates are present to capture the condensed liquid as it falls.

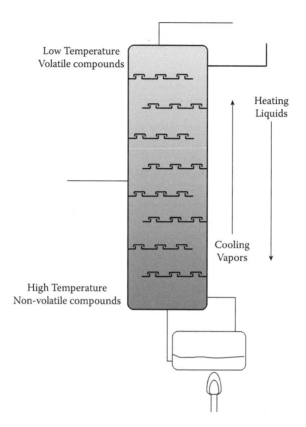

Figure 2.4 Diagram of a distillation tower. The coolest part of the tower is near the top, which is where very volatile compounds with low molecular weights are found and which is rich in vapor. The hottest part of the tower is at the bottom, which is where heavier compounds are found and which is mostly liquid.

Diesel fuel is a mixture of volatile organics that contain from 8 to 21 carbons and is distilled from crude oil between 200°C and 350°C at atmospheric pressure. Gasoline is a mixture of parrafins (hydrocarbons), naphthalenes (cyclic alkanes), and olefins (alkenes) that contain between 4 and 12 carbons per molecule, and is much more volatile than diesel fuel, meaning it can be distilled at a lower temperature. Both diesel fuel and gasoline that you purchase at the pump have several types of chemical additives added to the fuel to adjust its stability, resistance to compression, etc. These additives are predominantly organic molecules containing nitrogen, phosphorus, and oxygen, though aromatic molecules are also frequently used as additives. Aromatics are rings of carbon atoms where the type of C-to-C bonding exists as a hybrid of single and double bonds (see Appendix C for more details on organic chemistry). Because gasoline and diesel fuel are complex mixtures of many different compounds, we will focus the remainder of our combustion discussion upon gasoline and one of its major constituents, octane, for the sake of simplicity. However, a similar analysis can be applied to any type of fuel molecule.

Octane Number

Have you ever wondered what that octane number means at the pump, or why your car manufacturer says that you need to run premium gasoline? The octane number is a commonly accepted method for conveying the resistance of a fuel to self-detonation upon compression. In a gasoline IC engine, the fuel is not supposed to detonate until the spark plug fires. If, however, the engine compresses the fuel to the point that the fuel auto-detonates before the plug fires, this creates a problematic situation for the engine, often referred to as *engine knock*. A severe engine knock can cause permanent damage to critical engine components like valves and pistons, making it essential to avoid. In a lab, the octane number of a fuel is determined by burning the fuel at 600 rpm in an engine with a variable compression ratio. The goal is to identify the compression ratio where knocking is detected. The scientists and engineers then use the compression ratio where knocking occurred for the fuel to identify the mixture of isooctane and heptane with identical resistance to predetonation. The octane number assigned to the fuel is identical to the percentage of isooctane in the isooctane/heptane mixture with the same predetonation resistance. For example, if you put fuel in your tank with an octane number of 90, then whatever components are in that fuel have an equivalent resistance to predetonation as a mixture of 90% isooctane and 10% heptane. It is important to note that the octane number has nothing to do with the percentage of octane in your fuel. Remember, the fuel is a mixture of components all with their own octane numbers, and some compounds have better resistance to predetonation than isooctane, leading to octane numbers for some compounds of greater than 100. The take-home message is that the higher the compression ratio is in the car, the higher the octane number must be to prevent predetonation of fuel.

Why use gasoline and diesel as fuel sources? They have a number of extremely desirable properties for a portable fuel. First, they both contain molecules that are heavy enough to be liquids at atmospheric temperatures and pressures, yet they have vapor pressures that are low enough to start an engine under cold operating conditions in the winter. It is the vaporized fuel that is required for combustion to occur, and if the fuel does not have a flash point above the outdoor temperature, an engine will not start. They are also good fuel choices because they exist as liquids at typical atmospheric conditions, meaning they are dense enough that we can carry a lot of fuel in a relatively small storage container. Liquid fuels also offer some measure of safety versus compressed gases, the latter of which could easily detonate in an accident. As we will see later in this chapter, the "energy density" of a fuel, or the amount of energy contained in a unit volume, is also crucial for a fuel. Gasoline and diesel fuels both have very high energy densities. Gasoline and diesel fuels are also relatively stable in the sense that neither fuel tends to react with the gases in our atmosphere under normal temperatures and pressures. A fuel with a high energy density that reacts with air to form less energetic fuel molecules would cause the energy content of the fuel to potentially vary significantly between the initial fill-up and when the tank is nearly empty. A good fuel also needs to have a very fast reaction rate once the spark plug fires so that the fuel burns completely and quickly. To accelerate a car as quickly as possible or to have it travel at high speed, we need to repeat the reaction cycle many times in a small window of time. For example, a fuel that releases energy as slowly as burning wood would not be capable of rotating the crankshaft and wheels many times per second.

The combustion of gasoline and typical gasoline additives releases an enormous amount of potential energy stored in the fuel molecules. We can attempt to understand these energy releases by looking at thermochemical equations for the combustion of typical fuel components. Thermochemistry is the field of chemistry that is interested in energy and energy changes that take place during chemical reactions in the form of heat. A thermochemical equation differs from a standard type of chemical reaction in two important ways. One involves the stoichiometric coefficients. The coefficients in thermochemical equations refer to the number of moles of material required for that type of chemical conversion rather than the molecule-by-molecule basis in which most chemical reactions are written. The second is that thermochemical equations show the energy released or required as heat for that process listed as a number at the end of the equation. Since most combustion reactions in chemistry occur at either atmospheric or some other effectively constant pressure, by definition the heat released is equivalent to a thermodynamic state function called *enthalpy*, $\Delta_r H$. The enthalpy of combustion is a special type of reaction enthalpy defined to be the heat flow when 1 mole of a fuel is combusted with O_2 (g) as the oxidant.

State functions like enthalpy are functions that do not depend on path, or in other words, a change in that property can be determined simply by knowing the initial and final values of the property. Think about two people climbing to the top of a football stadium: if one climbs directly

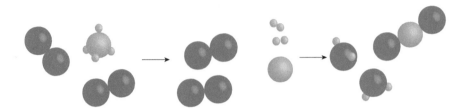

Figure 2.5 (See color insert.) Atom-scale view of the combustion of methane from the enthalpy-of-formation perspective. From a theoretical perspective, the reactants are viewed as breaking apart into their component elements; then the elements recombine to form new product compounds.

to the top up the stairs and the other runs around each row in the stadium before taking the next step upward, they both experience the same change in potential energy due to gravity if they start at the same elevation and end at the top row. However, if we examine the work that both individuals did to reach the top, these values will be quite different, since work is force times distance, and the distance the second person traveled is much longer. Person two will be much more tired. Therefore, energy is a state function that does not depend on path, while work (which we will discuss later in this chapter) is not a state function. State functions are powerful in thermodynamics because we can create or use any path we wish to determine or measure changes in their values, and this often lets us use tabulated parameters to determine the energy changes in unknown reactions.

In most cases, rather than measure energy changes in the laboratory, we take advantage of the fact that enthalpy is a state function and use enthalpies of formation for compounds to determine the overall enthalpy change of a reaction. A formation reaction by definition is one in which 1 mole of a material is created from its component elements in their most thermodynamically stable states, which are the states we find them in most often under a given set of conditions. In this model (Figure 2.5), we imagine that the reactants break apart into their constituent element's most stable forms (inverse formation reactions) and that the separated elements combine to form the products (forward formation reactions). This is not necessarily what really happens during the reaction, but because enthalpy is a state function, we can use this view to determine reaction enthalpies regardless of the exact reaction mechanism/pathway. Let's look at an example to understand the combustion of gasoline in your car by studying the combustion of octane, C_8H_{18}, via the enthalpy-of-formation model. The balanced overall chemical reaction for this process is:

$$C_8H_{18} + {}^{25}/_2O_2 \rightarrow 8CO_2 + 9H_2O$$

From the perspective of the enthalpy of formation, if we wish to determine the energy released by octane combustion, we first must break apart the reactants into their associated elements in what amounts to reverse enthalpy-of-formation reactions. At typical room temperature:

$$C_8H_{18} \rightarrow 8C \text{ (graphite)} + 9H_2 \text{ (g)} - \Delta_f H^0 = 250.1 \text{ kJ/mol}$$

$$O_2 \text{ (g)} \rightarrow O_2 \text{ (g)} - \Delta_f H^0 = 0 \text{ kJ/mol}$$

The negative sign in the $-\Delta_f H^0$ indicates that the energy change of a reverse formation reaction is the same magnitude, but opposite in sign, as that of a forward formation reaction. The reason for this equivalence will become clearer in the following section discussing chemical bonds. In the second reverse formation reaction, O_2 (g) is the most thermodynamically stable state of the element oxygen at room temperature, and the reactants and products look the same. Thus, there is no reaction, and the enthalpy change when no reaction occurs is zero. Now that we have dealt with the reactants, we can recombine the elements to form the products, which are forward formation reactions.

$$C \text{ (graphite)} + O_2 \text{ (g)} \rightarrow CO_2 \text{ (g)} \qquad \Delta_f H^0 = -393.5 \text{ kJ/mol}$$

$$H_2 \text{ (g)} + {}^1/_2 O_2 \text{ (g)} \rightarrow H_2O \text{ (g)} \qquad \Delta_f H^0 = -285.8 \text{ kJ/mol}$$

At this stage, we know how much energy it takes to form or dissociate 1 mole of each reactant and each product into their component elements. All that remains is to account for the number of moles of each component involved in the overall reactions and to add the results to obtain the overall reaction enthalpy. We make 8 moles of CO_2 (g) molecules, so we need 8×-393.5 kJ/mol to account for the carbon dioxide. Likewise, we need to account for 9 moles of H_2O molecules (9×-285.8 kJ/mol) and the breakup of 1 mole of octane molecules (250.1 kJ/mol). When we add these all together, we get:

$$(8 \times -393.5 \text{ kJ/mol}) + (9 \times -285.8 \text{ kJ/mol}) + 250.1 \text{ kJ/mol} = -5470 \text{ kJ/mol}$$

This is effectively identical to the experimentally measured enthalpy of combustion for liquid octane determined in a bomb calorimeter, −5470.5 kJ/mol. Thus, describing reaction energetics with an enthalpy-of-formation model is very powerful for examining unknown reactions. In addition, you can see one reason why octane makes a great fuel: it has a very high chemical potential energy, which means you can have a small fuel tank and make a car drive a very long distance. Analyzing reactions via an enthalpy-of-formation view is one example of Hess's law, which says that the overall change in the value of a state function during some process is the same regardless of the path used to get from your reactants to your products.

We can also use enthalpies of formation and combustion to understand why cars get poorer fuel economy with ethanol-inclusive fuels.
Since gasoline is a complex mixture of organic compounds (as noted
earlier), we will assume that gasoline is predominantly octane for this
exercise. In the previous paragraph, we determined that to burn 1 mole
of octane generates 5470 kJ of energy. Using the typical mass density of
gasoline (0.74 kg/L) and the principles of dimensional analysis, we can
determine the energy content per gallon of a pure octane gasoline, called
an *energy density*.

$$\frac{5470 \text{ kJ}}{\text{mol}} \times \frac{1 \text{ mol } C_8H_{18}}{114.224 \text{ g}} \times \frac{1000 \text{ g}}{\text{kg}} \times \frac{0.740 \text{ kg}}{\text{L}} \times \frac{3.79 \text{ L}}{\text{gal}} = 134,000 \text{ kJ/gal}$$

Now, we can use our concepts of Hess's law and thermochemical
equations to determine the energy released by burning 1 mole of ethanol,
and then use the idea of mass density and molecular mass to determine
the energy density of ethanol.

$$C_2H_5OH + 3O_2 \rightarrow 2CO_2 + 3H_2O \qquad \Delta_c H^0 = -1366.8 \text{ kJ/mol}$$

$$\frac{1367 \text{ kJ}}{\text{mol}} \times \frac{1 \text{ mol } C_2H_5OH}{46.068 \text{ g}} \times \frac{1000 \text{ g}}{\text{kg}} \times \frac{0.790 \text{ kg}}{\text{L}} \times \frac{3.79 \text{ L}}{\text{gal}} = 88,800 \text{ kJ/gal}$$

Note that there is substantially less energy per gallon of ethanol
than there is per gallon of octane. Now let's say your drive to work
every day on the interstate highway requires 200,000 kJ of energy.
This energy is used to accelerate the vehicle to the cruising speed and
includes energy lost to inefficiencies in the engine and drivetrain,
frictional forces to maintain that cruising speed over the distance to
work, energy converted to electrical energy by the alternator, etc. If you
have a car that contains 100% octane as a fuel, we can do some simple
math to show that it requires you burn 1.49 gallons of gasoline to drive
this distance.

$$\frac{200,000 \text{ kJ}}{134,000 \text{ kJ/gal}} = 1.49 \text{ gal}$$

Let's compute the volume of fuel required for the same drive if your
tank is 90 vol% octane and 10 vol% ethanol. Then the average energy content of one gallon of fuel is going to be the weighted average of the pure
ethanol and pure octane components. You likely have experience calcu-
lating weighted averages from general chemistry when you calculated
the average molecular mass of elements in the periodic table. Based on
our energy density calculations, we hypothesize that our trip will require

more fuel in the ethanol-enriched gasoline to go the same distance. The calculation supports our hypothesis:

$$\frac{200,000 \text{ kJ}}{[(0.9 \times 134,000 \text{ kJ}) + (0.1 \times 88,800 \text{ kJ})]/\text{gal}} = 1.54 \text{ gal}$$

The situation is even worse for an E-85 vehicle burning 85 vol% ethanol, where the same type of calculation shows that you need to burn ≈2.1 gal of this fuel to cover the same distance (do this example calculation on your own). Even though real gasoline is a complex mixture of chemicals rather than pure octane, the key result of this simplified thermochemical analysis is that including ethanol in gasoline reduces your fuel economy and that the fuel economy gets worse as the percentage of ethanol in the fuel increases.

Ethanol and Our Energy Future

Despite our thermochemical results, many gas stations sell ethanol-enriched fuels and the federal government provides subsidies for ethanol production facilities at the writing of this text. Clearly, including ethanol in fuels must provide some advantages. One is that ethanol burns at a lower temperature than gasoline or diesel fuel, which reduces the production of soot and NO_x gases (NO and NO_2) that are generated primarily via reactions of N_2 and oxygen that only occur at very hot temperatures, such as those in the cylinder gases and exhaust system. Ethanol has an octane number of 113, and thus, including ethanol also provides a chemical means to tune the octane number of a fuel. In addition, at least in principle, the drop in fuel economy calculated here is not so severe in practice, since ethanol burns more efficiently in an engine because it is an oxygenated fuel. Oxygenated fuels are molecules that contain their own oxygen, which helps to ensure there is enough oxygen in the reaction chamber to achieve complete combustion. This oxygenation also contributes toward reports that ethanol is a less polluting fuel, although modern catalytic converters are often capable of cleaning pure gasoline exhaust vapors so well that they largely negate the cleaner-burning advantages of oxygenated fuels. Diluting gasoline with locally produced ethanol also provides a strategic geopolitical advantage in the sense that it reduces the need to import fuel from other nations, improving our national energy security. And of course, ethanol produced from crops and crop wastes is a renewable fuel that consumes nearly all of the CO_2 it generates during combustion in the next growing cycle—a so-called zero-net CO_2 fuel.

To determine whether ethanol really provides any substantial energy advantage or reduction in greenhouse gas production requires a complete life-cycle analysis. Life-cycle analyses are complicated endeavors that require accounting for all energy inputs, outputs, and economic effects of a fuel from cradle to grave. They often involve many educated assumptions that affect their results. Most models show that the ethanol feedstock (corn, sugarcane, etc.) is an important factor in whether crop ethanol is an energy advantage. Choice of feedstock certainly has an additional impact on the global food supply. Modeling energy economics is well beyond the scope of this text, but clearly ethanol in fuels is a hot topic that highlights the interplay of science, society, values, politics, and economics. For a more detailed discussion of alcohol biofuels, the interested reader is referred to the excellent discussion in the 5th edition of Baird and Cann's Environmental Chemistry.*

2.3   Bonds and bond energy

Chemistry Concepts: types of bonding, thermochemistry, coulombic attraction, Hess's law

Expected Learning Outcomes:
- Explain where chemical potential energy is stored
- Understand that breaking bonds takes energy and making bonds makes energy
- Draw a typical potential-energy diagram for formation of a covalent bond
- Explain how Hess's law allows the bond-breaking/making model to explain reaction enthalpies

In the previous section, we discussed reaction enthalpies and examined the combustion of octane using an enthalpy-of-formation model. Recall that the enthalpies of formation represent the combination of elements in their most stable states to form compounds. Elements like carbon, oxygen, nitrogen, and phosphorus combine in organic compounds by sharing electrons with other C, O, N, P, or H. This sharing of electrons is known as a chemical bond (in particular, a covalent bond), and the making and breaking of chemical bonds is how energy is stored or released in reactions as chemical potential energy.

The common elements in organic compounds are willing to share electrons to form covalent bonds because forming such bonds leads to a lower

* C. Baird and M. Cann, *Environmental Chemistry*, 5th ed. (New York: W. H. Freeman, 2012).

energy state, making the bonded atoms more thermodynamically stable than two infinitely separated atoms. We can understand the formation of covalent bonds based on electrostatic (or coulombic) interactions between positive charges, that is, the nuclei, and the negative charges, that is, the electrons. Qualitatively, coulombic interactions tell us that opposite charges attract one another and that like charges repel one another. For the sake of simplicity, let us consider the approach of two hydrogen atoms toward one another. Each atom has a single positive charge at its nucleus and a single negatively charged electron. When the two atoms are very far away from one another, there is effectively no net interaction energy between them (Figure 2.6). However, as they approach one another, each electron begins to notice the positive charge of the other atom's nucleus. The electrons begin to be attracted by both nuclei at this stage. It takes energy to keep the electrons away from the second nucleus, and thus the total potential energy drops as we let the atoms get closer together. At some point, if we attempt to force the atoms closer still, the two nuclei begin to feel each other's positive charge. Since like charges repel, it takes energy to get the nuclei closer together, and the energy of the system increases to well above the infinite separation value as the internuclear distance approaches small values ($\approx$<1 Å, or 1×10^{-10} m). Essentially, the attraction of both electrons to both nuclei is opposed by the repulsion between the two dense positive charges of the nuclei. This competition causes a minimum in the potential energy curve at some particular internuclear distance. That distance is called the *bond distance*, and the extra

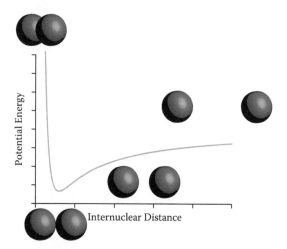

Figure 2.6 A potential-energy diagram for the formation of a covalent bond. The distance that yields the lowest potential energy fixes the equilibrium covalent bond length.

energetic stabilization by sharing the electrons defines the covalent bond. It requires energy to separate the atoms to a larger distance, and it takes energy to force them closer together. Since the energy minimum occurs at a negative value of the total energy with respect to infinitely separated hydrogen atoms, forming a bond always releases energy. And if forming bonds releases energy, then breaking bonds apart always requires an addition of energy.

The source of energy in the previously noted formation reactions and reverse formation reactions is the breaking and making of chemical bonds. Formation reactions confirm several insights from our coulombic analysis of bond making and breaking, one of which is that forming bonds releases energy (thus the negative sign on all formation enthalpies) and that breaking bonds requires energy (thus the positive sign on all reverse formation reactions). The bond view also explains why the energy released by making a bond is identical to the energy required to break that bond. Breaking bonds requires that we move up the right-hand slope in Figure 2.6 to infinite separation as we separate the atoms on the reactant side, while making bonds requires that we bring atoms from a large separation to the minimum energy. In terms of our potential energy diagram, these two processes traverse the same path but in opposite directions, meaning that it takes the same energy to break specific kinds of bonds as is released by making specific kinds of bonds.

Although our previous discussion revolved around two hydrogen atoms, the same principle can be used to explain the formation of any covalent bond between any two atoms. The electrons that participate in covalent bond formation are always the outermost electrons in an atom, called the valence electrons. The coulombic attractions of the valence electrons for each nucleus in a compound and the repulsions between the nuclei give rise to the sharing of the valence electrons. These principles also guide the distance between atoms, their locations in space, and the types of chemistry that the molecules can undergo. Identifying the locations of atoms and electrons and the associated energies are at the heart of molecular modeling, which plays an increasingly important role in understanding the details of chemical processes.

The idea of bond energies also gives us a new path to consider when computing the enthalpy of a reaction—the bond energy pathway. When viewing the energy changes in a chemical reaction in this way, we imagine taking the reactants and breaking them apart into individual atoms, then allowing the atoms to recombine to form the products. This is different than the formation enthalpy model. In the formation view, intermediate oxygen exists as O_2 (g) for reactions at typical atmospheric temperature, while in the bond energy view, oxygen exists as atomic O in the intermediate state. We must put in enough energy to break all of the bonds present in the reactants, and then we get energy back for every bond we form in the products. Let's consider the combustion of ethanol from a bond-energy perspective:

$$C_2H_5OH + 3O_2 \rightarrow 2CO_2 + 3H_2O \qquad \Delta_cH^0 = -1366.8 \text{ kJ/mol}$$

Ethanol contains one C–C bond, five C–H bonds, one C–O bond, and one O–H bond. Each oxygen atom contains one O=O, so we must break three O=O bonds to complete our bond breaking on the reactant side of our reaction. Typical bond energies for these bonds are 347 kJ/mol for C–C, 413 kJ/mol for C–H, 358 kJ/mol for C–O, 467 kJ/mol for O–H, and 495 kJ/mol for O=O, giving us a total energy expense to break all the bonds on the reactant side of:

$$347 \, \frac{\text{kJ}}{\text{mol}} + \left(5 \times 413 \, \frac{\text{kJ}}{\text{mol}}\right) + 358 \, \frac{\text{kJ}}{\text{mol}} + 467 \, \frac{\text{kJ}}{\text{mol}} + \left(3 \times 495 \, \frac{\text{kJ}}{\text{mol}}\right) = 4722 \, \frac{\text{kJ}}{\text{mol}}$$

On the product side, we form four C=O and six O–H bonds when we recombine the elements to form the products. With average bond energies of 799 kJ/mol for C=O and 467 kJ/mol for O–H, the same type of calculation yields a total energy release from new bonds forming of 5998 kJ/mol. Thus, we arrive at an enthalpy of combustion for ethanol of –1276 kJ/mol, very close to our earlier answer. We note that bond energies are often reported as average bond energies, with the understanding that bond energies vary to some degree, depending on the overall structure of the molecule. It is likely that our discrepancy is a result of the tabulated average bond energies not reflecting the true reality of bonding in ethanol. Table 2.2 provides average bond energies useful to thermodynamic calculations of fuel combustion.

Table 2.2 Bond Energies Useful in Fuel Combustion Calculations

Single bonds		Double bonds	
Bond type	Avg. Bond energy	Bond type	Avg. Bond energy
C–C	347 kJ/mol	C=C	620 kJ/mol
C–H	414 kJ/mol	O=O	498.7 kJ/mol
C–O	351 kJ/mol	C=O	745 kJ/mol
C–N	276 kJ/mol	C=O (CO_2)	799 kJ/mol
C–S	255 kJ/mol	N=O	607 kJ/mol
O–H	460 kJ/mol	S=O	469 kJ/mol
N–H	393 kJ/mol		
N–O	176 kJ/mol		

Source: Adapted from R. Chang, *General Chemistry: The Essential Concepts,* 5th ed. (New York: McGraw-Hill, 2008).

2.4   *Heat, work, and engine efficiency*

Chemistry Concepts: thermochemistry, first law of thermodynamics, state functions

Expected Learning Outcomes:
- Write and explain the first law of thermodynamics
- Understand the symbiotic relationship of heat and work and how these relate to engine efficiency
- Explain the difference between heat and work

Not all of the energy released in the combustion reaction of an IC engine is converted to the kinetic/mechanical energy that moves the vehicle. The reason for this is intimately tied to the first law of thermodynamics, which says that the internal energy of an isolated system is constant. An isolated system is one that is forbidden from exchanging matter or energy with the rest of the universe. The internal energy of a system is the sum of all the energies within the system, including the chemical, translational, vibrational, and rotational energy of every molecule of every compound within the system. The chances of encountering a truly isolated system here on Earth are approximately zero, and an internal combustion engine certainly does not qualify as one. The combustion gases are in contact with the engine and exit via the exhaust system, which means that heat energy is exchanged with the surroundings, and the engine also exchanges matter with the surroundings in every combustion cycle. So what exactly happens to the energy released during combustion and how much is harvested for moving the vehicle?

To help answer this question, we must view the first law of thermo-dynamics in another way: Any change in the internal energy of a general system is exactly equal to the energy evolved as heat (q) and the energy used to do work (w):

$$\Delta E = q + w$$

Heat is the transfer of energy via the random motions of molecules, and is typically observable as a change in the temperature of an object. Work is the transfer of energy via the organized motion of molecules. For example, if you push a box across the floor, you are moving all the molecules in the box in an organized fashion, and it takes work on your part. Your body already recognizes this: you will get tired if you push a large enough box long enough, a sure sign of doing work. Likewise, the expansion of a gas in an engine cylinder pushes down all the molecules in the piston in an organized fashion, meaning it takes work to move the piston, turn the crankshaft, etc.

The first law of thermodynamics shows that heat and work have a symbiotic relationship. To best grasp this concept, think about trying to increase the temperature of 100 g of water by 10°C. To bring about the required change in internal energy, you could provide all that energy as heat, as you might do on your stove. But the first law shows that you could also accomplish this change in internal energy by not heating the water at all, provided that you do enough work on the water. You can do this using a pump, mixer, etc. If you own a compressor or an aquarium pump, you can feel the increase in temperature of the outlet gas after the device has been running for a small amount of time. This observable temperature change in your aquarium pump is in large part a result of the work being done by the pump/compressor to move the air. Likewise, when your body does work by pushing a heavy box, you also get hot because your body is not 100% efficient at converting energy into work. The key result is that heat and work can be converted into one another.

The simplest type of work typically discussed in a chemical thermodynamics course is the work of expansion for gas systems, where the work of expansion is related to the integral of pressure times the change in volume:

$$w = \int -p\, dV$$

We also know that the ideal-gas equation tells us that the change in the product of pressure and volume is directly proportional to the change in the product of temperature and the moles of gas:

$$\Delta(pV) = R\Delta(nT)$$

By combining these two equations, we see that in a very general case, the amount of expansion work that the engine can do is directly related to the temperature change of the gas during combustion. However, we also know that an engine loses heat from this combustion process to the materials that make up the engine itself and to the hot exhaust gases, which means that not all of the energy released is used to do expansion work.

In fact, it can be shown that the first law of thermodynamics has significant implications in terms of engine efficiency, and that the maximum theoretical efficiency is related to the difference between the upper and lower operating temperatures of the engine. One way to think about efficiency is to consider how much of the energy released in the combustion process is used to do the work of moving the vehicle. A 100% efficient engine in this definition would convert all of the chemical energy released as heat into work that moves the car. Practically, the maximum efficiency is limited by materials chemistry. The theoretical upper operating temperature of any engine will be the melting point of the material used to construct the engine, and typically the actual upper operating temperature is below that value. Internal combustion engines can be considered to be heat engines, and the ideal engine efficiency for any heat engine can be calculated using the Carnot cycle (for a discussion, see an upper-level physical chemistry text). When one performs such a calculation for an iron/steel engine block—assuming that the lower operating temperature is equal to the operating coolant temperature in a typical engine ($\approx$220°F)—the upper limit to efficiency is 75%. In reality, the lower operating temperature of the combustion gases in the cylinder head is much higher than the coolant temperature, leading to maximum ideal efficiencies in the 30% range.

Practically, the efficiency of an engine is substantially lower than the ideal value because (a) the engine does not operate in service at the theoretical upper operating temperature and (b) there are losses associated with other kinds of work that are done before the flywheel. It takes work to overcome the friction between the piston heads and cylinder walls, to turn all the accessories on the engine, to pump the air in and then out the exhaust, etc. Engines also have an optimum power output under a particular set of operating conditions (as will be discussed in the following section) and typically operate away from that design optimum. Real four-stroke engines are better represented thermodynamically using the Otto cycle (see a more advanced text for details), which shows that ideal efficiency is related to the compression ratio (the change in volume from bottom dead center to top dead center) and the ratio between the constant-pressure and constant-volume heat capacities of the gas mixture. For a typical engine with a 10:1 compression ratio, the average Otto efficiency is roughly 45% for a frictionless engine, and is practically in the 30%–35% range when one accounts for frictional losses. Other losses to friction and heat in the crankshaft, flywheel, camshafts, and valves, motion of the pistons in the other three strokes, pumps, and accessories all reduce the efficiency measured at the flywheel to the 18%–22% range. More detailed discussions of heat engines and engine efficiency can be found in more advanced engineering and physics texts.

Hybrid Vehicles and Regenerative Systems

One approach that some manufacturers are taking to improve the overall efficiency of cars is to incorporate regenerative systems with dual IC/electric drivetrains. Vehicles of this type are often called hybrids, referring to the mix of electric motors and combustion-based power plants that do the work of moving the vehicle. Hybrid vehicles recover energy that is otherwise lost and use it to charge batteries. For example, conventional braking systems simply dissipate the energy of the moving vehicle by converting it to heat that is then lost to the atmosphere. However, if electric motors are attached to a set of wheels, the motors can be used as generators during braking. The resistance to rotation when the motor is in generator mode helps to slow the vehicle and, at the same time, converts the kinetic energy of the wheels into electrical energy that can be stored in a battery. These batteries then power the electric motors that help to drive the vehicle at low speeds. Since the system recovers energy that would otherwise be lost during braking, the overall efficiency of the vehicle improves. Some manufacturers are currently targeting automatic transmissions for regenerative electricity generation as well, using the rotation of the clutch packs during deceleration as generators to harvest kinetic energy that is otherwise lost. When combined with features like an engine that shuts down when the car is stopped, hybrid vehicles with regenerative systems can substantially improve fuel economy in stop-and-go driving scenarios. As a side note, at this time, hybrids provide little improvement in efficiency or fuel economy at highway speeds, where the power requirements and lack of any regenerative battery charging necessitate the engine handling nearly all the load.

2.5   Power and the fuel source

Chemistry Concepts: thermochemistry, first law of thermodynamics, torque
Expected Learning Outcomes:
  • Understand how engine power and torque ratings relate to chemical energy
  • Convert between energy, horsepower, and torque

As noted in the previous section, there are several factors that limit engine efficiency, many of which are difficult if not impossible to overcome. However, there are some factors that we can control to increase the performance of an engine, namely manipulating the combustion chemistry either through fuel selection or oxidant type, amount, etc. This section and the following section take a look at these two types of manipulation in greater detail.

Before discussing how alternative fuels impact engine performance, it is necessary to discuss the criteria conveyed to consumers about the energy output of an engine and the work that the engine can do. These are typically provided as peak horsepower and torque ratings for an engine. Torque in physics is the tendency of a force to rotate an object about an axis, and is related to work for rotating a body:

$$w = \int_{\theta_1}^{\theta_2} \tau d\theta$$

This equation can be used to calculate the work required to rotate the flywheel, driveshafts, wheels, etc. Here, θ_1 and θ_2 are the initial and final angles of rotation with respect to the reference position. Torque is important for an internal combustion engine because the portion of chemical energy harvested to do the work of moving the car causes rotation of the crankshaft, flywheel, driveshafts, axles, and even the wheels. Torque effectively tells us how easily the engine can rotate these components, and it relates directly to how strongly you "feel" a car accelerating. Power is defined as the energy released per unit time, and therefore if we know the amount of time it takes for a single cycle and the engine power, we can calculate the energy release (or any permutation of the three quantities). For example, completion of a full combustion cycle requires two revolutions of the crankshaft. If you know the amount of fuel injected during one combustion cycle, you can calculate the energy released during combustion. If you know the engine rpm (revolutions per minute), you can calculate the time required for two full revolutions by taking its reciprocal and multiplying by two, giving the following formula for relating enthalpy of a cycle to horsepower:

$$\text{hp} = n \times \Delta_c H \times \frac{\text{rpm}}{2} \times \frac{\text{min}}{60 \text{ s}} \times \frac{1 \text{ hp}}{746 \text{ J/s}}$$

where n is the number of moles of fuel combusted in a cycle (total over all the cylinders), the enthalpy term is the enthalpy of combustion for the fuel, and rpm is the engine rotational velocity. Power and torque are also directly related: The power of an engine is by definition the angular velocity of the crankshaft multiplied by the torque. Thus, energy, power, and torque are all related to one another and easily converted to one another. Also, it is apparent that both horsepower and torque are functions of the crankshaft rotational velocity. In other words, the power and torque output at any given moment depend on the engine rpm. The peak horsepower and peak torque are the maximum values of horsepower and torque observed between the engine idle and maximum rpm (redline), and their values are typically reported with the rpm where they hit their maximum. Ideally, an engine reaches its maximum torque very quickly and has a flat, stable value over the entire rpm range. The power peak always occurs before the maximum rpm in real engines.

To further illustrate these relationships, we will work through a short example involving the engine of a top-fuel dragster. At the writing of this text, top-fuel dragsters run a 500-in.[3] (8.1 L) V8 engine that maxes out around 8400 rpm. Though these cars run a 90% nitromethane/10% methanol fuel during a run down the drag strip, the engines are always started and stopped while burning conventional gasoline. If we assume that the oxygen in air provides the limiting reagent and that the inlet air temperature is equal to 25°C, then we can compute the number of moles of air consumed in one combustion cycle by the entire engine using the ideal-gas law and Dalton's law of partial pressures, which essentially states that the pressure in a mixture of gases attributed to one particular component of that mixture is related to the mole fraction of that component gas. Since we are interested only in the oxygen contained in air, the relevant mathematical statement of Dalton's law is:

$$p_{O_2} = X_{O_2} p_{atm}$$

where p_{O_2} is the partial pressure of O_2, X_{O_2} is the mole fraction of oxygen in the atmosphere, and p_{atm} is the total atmospheric pressure. Assuming that the ambient pressure is 14.7 psi (1 atm), and for simplicity that there are no pressure variations arising from any component in the engine's intake system, Dalton's law tells us that we will have a pressure of O_2 equal to 0.21 atm in our combustion chamber, giving 0.0695 mol as the total number of moles of O_2 per cycle.

$$p_{O_2} = 0.21 \frac{\text{mol } O_2}{\text{mol air}} (1 \text{ atm}) = 0.21 \text{ atm}$$

$$n = \frac{PV}{RT} = \frac{1 \text{ atm} \times 8.1 \text{ L}}{(0.0821 \text{ L atm/mol} \cdot \text{K}) \times 298 \text{ K}} = 0.331 \text{ mol} \times 0.21 \frac{\text{mol } O_2}{\text{mol air}}$$

$$= 0.0695 \text{ mol}$$

Looking back at our enthalpy of combustion information for octane, we see that we get 5470 kJ of energy for every 25/2 moles of oxygen gas that are consumed. Using simple concepts in stoichiometry and dimensional analysis, we find that for this amount of oxygen, the engine generates ≈30 kJ of energy per cycle.

$$0.0695 \text{ mol} \times \frac{5,470,000 \text{ J}}{\frac{25}{2} \text{ mol}} = 30,424 \text{ J}$$

If the engine is fed fuel and run at the redline of 8400 rpm, then the horsepower created by this engine under naturally aspirated conditions will be:

$$30,424 \text{ J} \times \frac{8,400 \text{ revolutions}}{\text{min}} \times \frac{\text{min}}{60 \text{ s}} \times \frac{\text{cycle}}{2 \text{ revolutions}} \times \frac{1 \text{ hp}}{746 \text{ J/s}} = 2,854 \text{ hp}$$

Of course, this number assumes that gasoline is 100% octane and that the engine is 100% efficient at converting the heat energy into work, which we know to be untrue from the previous discussion. If we assume a more reasonable 20% efficiency for the engine, this number is a realistic 570 hp at the flywheel. We can also determine the torque at this rpm level:

$$570 \text{ hp} \times \frac{33,000 \text{ ft} \cdot \text{lb/min}}{\text{hp}} \times \frac{1}{2\pi \times 8,400 \text{ rpm}} = 356 \text{ ft} \cdot \text{lb}$$

However, as noted earlier, top-fuel dragsters do not run on conventional gasoline, but rather on a mixture of nitromethane and methanol to generate in excess of 6000 peak hp. This is a classic example of manipulating the combustion chemistry to generate additional performance and power. To understand why burning nitromethane generates so much more power, we start with the combustion reaction for nitromethane.

$$CH_3NO_2 + {}^5/_2O_2 \rightarrow CO_2 + {}^3/_2H_2O + NO \qquad \Delta_cH = -747.6 \text{ kJ/mol}$$

A quick examination of this shows that you generate substantially less energy per mole of nitromethane than you do burning octane. However, recall that our previous analysis assumed that oxygen was the limiting reagent in the combustion reaction. Thus, we have the same amount of air as in the octane combustion case, but substantially more favorable stoichiometry: essentially, we can burn five times more nitromethane with the same amount of oxygen based on the stoichiometry alone.

$$0.0695 \text{ mol} \times \frac{747,600 \text{ J}}{\frac{5}{2} \text{mol}} = 20,783 \text{ J}$$

Although this is still a lower-than-reported theoretical horsepower (1950 hp), this comparison assumes that gasoline is 100% octane, and the calculation we just performed does not account for the fact that nitromethane has a much higher density than gasoline (1.137 g/mL versus ≈0.75 g/mL), nor does it account for the extra oxygen that comes from the methanol component of the racing fuel. In the end, the difference in energy density, reaction stoichiometry, and several other factors show that you can realistically make up to 2.3 times the power by burning nitromethane in place of octane in an IC engine. A real top-fuel dragster engine makes additional power as a result of efficiency increases due to the supercharger (see the following section) and an upper operating temperature closer to the theoretical limit for the engine materials compared to a typical street vehicle. The enthalpy of formation, energy density, and other analyses discussed thus far in this chapter can also be used to identify fuel alternatives for greater power generation and explain why top-fuel dragsters and funny cars make enormous amounts of horse-power by burning nitromethane, which is, on paper, a poorer fuel source than gasoline.

Diesel Racing Engines

In recent years, many automobile manufacturers have raced diesel-engine vehicles in high-profile venues, such as the Audi R10 TDI, that dominated the 24 hours of Le Mans. Diesel engines are more efficient than gasoline engines, which means that diesel vehicles will have a fuel-economy advantage and perhaps need to stop less frequently in the pits for refueling, a benefit in a long race. However, diesel engines always oper-ate at lower rpm than their gasoline counterparts because they require heavier components to withstand the stresses in the combustion chambers due to the higher pressures and very high compression ratios. Thus, for a given energy release, performing calculations like those in this section show that diesels will produce less horsepower than a gasoline combus-tion engine of similar displacement, since the rpm multiplier will be smaller. So why do teams race diesels? The answer is that diesel engines make up for this power disadvantage by providing a significant torque advantage. Nearly all mod-ern diesel engines are designed to produce large amounts of torque by building the engines with very long stroke lengths. The longer connecting rod between piston and crankshaft gives the diesel piston more leverage, making it easier to rotate the crankshaft, and the ability to rotate a body is the definition of torque. At the same time, the pressure in a diesel cylinder is much higher than that in a gasoline engine as a result of the higher compression ratios and the method of introducing fuel into the combustion chamber. Since diesel engines spray fuel directly into the heated compressed air rather than introduc-ing a single fuel/air charge like a gasoline engine, the diesel engine can inject fuel in a manner that permits combustion to occur over a larger portion of the downstroke, generating addi-tional pressure compared to a gasoline engine. These higher pressures impart a larger impetus for the piston to move, and together, these factors result in diesel engines making a large amount of torque. This high torque helps them accelerate quickly out of corners, and with proper gearing in the trans-mission, a diesel race car can be quite competitive, even with lower peak horsepower.

2.6   Turbochargers and superchargers

Chemistry Concepts: gas laws, mass transport, limiting reagents, thermodynamics

Expected Learning Outcomes:

- Use thermodynamics and the ideal-gas law to explain how turbochargers and superchargers allow an engine to generate additional power and/or improve fuel economy
- Use your understanding of heat transfer to explain why intercoolers are useful devices in forced-induction systems

As discussed earlier in this chapter, an internal combustion engine or a diesel engine generates power by capturing a portion of the energy released in a combustion reaction to do expansion work. As the gas-phase reaction products expand post-ignition, they push down on the piston, which rotates the crankshaft and flywheel, and the kinetic energy of these components is transferred to the wheels via the transmission. The energy that survives all of these transfers is used to do the work of moving the vehicle. Even though much of the energy released during combustion is lost to heat, friction, and drag (the resistive force exerted by a fluid on a moving solid) in this sequence of events, this system is effective because the combustion-type oxidation of the carbon in the fuel is a very highly exothermic process that also generates a net increase in the number of gas-phase molecules.

As with any chemical reaction, the combustion of fuel involves a limiting reagent. Recall that the combustion chamber in an internal combustion engine is really a highly specialized reaction vessel where the fuel is mixed with the oxidant (O_2) and ignited via an electrical discharge that is used to overcome the activation-energy barrier to combustion. We have nearly infinite control over the rate of fuel delivery to this vessel (within the limits of the fuel pump properties and other components in the delivery system) and complete control over the spark system, making the limiting reactant in most cases the oxygen in the combustion chamber. Thus, improving the delivery of our limiting reagent, O_2, is a great way to improve an engine's power and efficiency.

In a naturally aspirated four-cycle engine, the driving force for bringing O_2 into the combustion chamber is simply the vacuum generated by the downward stroke of the pistons on the intake step. In an ideal case, the motion of the pistons in one full cycle will draw in a volume of gas equal to the displacement of the engine. In reality, frictional losses and other effects cause a normally aspirated engine to draw less than its full displacement of air into the combustion chamber. The ratio between the amount of air the engine truly draws and the theoretical amount of air that the engine can draw is known as the volumetric efficiency, η_V, which is

Figure 2.7 Cutaway view of a conventional turbocharger, showing an example of a centrifugal compressor on the right-hand side of the image. On the left are the turbine blades driven by the engine exhaust, which cause the impeller on the right to rotate and accelerate inlet gas (drawn in at the rightmost opening) into the diffuser chamber (round holes parallel to plane of image). (Image adapted from NASA. Turbocharger cutaway image was released to the public domain with no copyright by NASA.)

always less than the ideal value of one. This sub-ideal volumetric efficiency combined with the relative low abundance of O_2 in air (21 atom%) represent a significant limitation when it comes to efficient power generation in combustion-based engines.

Turbochargers and superchargers are centrifugal compressors that force more air into the engine, overcoming the inherent volumetric efficiency limitations and leading to a substantial increase in the energy released in an engine cycle. A centrifugal compressor is a device that uses impellers to impart additional kinetic energy to a gas, in this case, the inlet air (Figure 2.7). The rapidly rotating impeller blades force the gas to the outside of the compressor housing, increasing the kinetic energy and causing the gas to flow in a spiral pattern. The increase in velocity can be found using one form of Euler's fluid dynamics equation:

$$Ws = u_{out}C_{\theta,out} - u_{in}C_{\theta,in}$$

where W_S is the shaft input power, u represents the blade tip velocity, and C_θ represents the tangential velocity of gas leaving the impeller blades at the inlet (in) and outlet (out). As the accelerated gas exits the impeller stage, it enters a device called a diffuser, which slows the gas molecules back down. The net effect of the diffuser is to convert the kinetic energy added by the impeller into potential energy in the form of increased gas pressure. This is essentially a reverse application of the Bernoulli principle, which states that an increase in the velocity of a gas occurs concurrently with a decrease in the gas pressure. While both superchargers and turbochargers operate on these principles, they differ in the driving force that rotates the impeller blades. Superchargers are driven by the pulley system on the engine (taking some power to make additional power—in fact, it can take up to 1000 hp from the engine to turn the supercharger on top-fuel dragsters), while turbochargers are driven by the heat and flow of the exhaust gases (recovering energy that is otherwise lost). Thus, turbochargers could be considered "energy recovery systems" and lead to greater improvements in engine efficiency.

Let's take a look at the combustion process again in a typical engine and in a turbocharged version of that engine to develop a deeper understanding of the role played by forced-induction systems. In this analysis, we will use several concepts typically presented in a general chemistry course: reaction enthalpy, stoichiometry, and the ideal-gas law.

If we make the simple assumption that all of the energy harnessed by the engine arises from combustion of pure octane, then our balanced thermochemical reaction at a temperature of 298 K is identical to that of octane combustion in Section 2.2:

$$C_8H_{18} + {}^{25}/_2O_2 \rightarrow 8CO_2 + 9H_2O \qquad \Delta_cH^0 = -5470 \text{ kJ/mol}$$

Let's use the 2.5-L boxer engine in a 2005 Subaru Outback[*] as our test case and assume that this engine has a volumetric efficiency of 80%, a typical value for a normally aspirated internal combustion engine. If we also assume that our fuel pump can provide as much octane as needed stoichiometrically for our reaction, then to determine the energy yield of a single engine cycle, we only need to calculate the amount of O_2 that can be delivered in that cycle, as we did in the previous section. If we assume that the intake temperature of the air is a typical atmospheric temperature of 20°C (293 K), that the reaction enthalpy at this temperature does not differ significantly from the tabulated enthalpies at 298 K, and that the inlet air temperature is not affected by any component of the intake system, then we can combine the ideal-gas law and partial pressure of O_2 used earlier (modified using the volumetric efficiency of the engine) to determine the moles of O_2 available for combustion in one cycle and the energy released in a single engine cycle.

---

[*] *Subaru Outback Literature Pack* (Subaru of America, 2006).

$$n_{O_2} = \frac{p_{O_2} V_{cycle} \eta_V}{RT}$$

$$n_{O_2} = \frac{0.21\, atm \times 2.5\, L \times 0.80}{0.0821 \dfrac{L\, atm}{mol\, K} \times 290\, K}$$

$$n_{O_2} = 0.017\, mol\, O_2$$

$$\Delta_c H^0_{1\, cycle} = -5470 \frac{kJ}{mol\, C_8H_{18}} \times \frac{1\, mol\, C_8H_{18}}{\dfrac{25}{2}\, mol\, O_2} \times 0.017\, mol\, O_2 = -7.4\, kJ$$

Now let's examine the 2005 Subaru Outback 2.5 XT, which in principle has the exact same engine equipped with a turbocharger that produces 13.5 psi of boost. This means that the turbocharger will add 13.5 psi of pressure to the initial atmospheric pressure, which in our example is 1 atm or 14.7 psi, giving a total intake pressure of 28.2 psi or 1.92 atm of air. If we use the same conditions as our normally aspirated example and make the same assumptions (again, for the sake of simplicity), we should be able to determine the energy output of the turbocharged version of the engine. Our hypothesis is that the energy output will be substantially increased.

$$p_{O_2} = 0.21 \frac{mol\, O_2}{mol\, air}(1.92\, atm) = 0.40\, atm$$

$$n_{O_2} = \frac{0.40\, atm \times 2.5\, L \times 0.80}{0.0821 \dfrac{L\, atm}{mol\, K} \times 290\, K} = 0.034\, mol\, O_2$$

$$\Delta_c H^0_{1\, cycle} = -5400 \frac{kJ}{mol\, C_8H_{18}} \times \frac{1\, mol\, C_8H_{18}}{\dfrac{25}{2}\, mol\, O_2} \times 0.034\, mol\, O_2 = -14\, kJ$$

This is a ≈92% increase in the energy output from a single cycle, which correlates well with the pressure boost from the turbocharger and supports our hypothesis.

Of course, the actual output of the turbocharged engine does not effectively double. According to the published engine data from Subaru, the normally aspirated 2005 Outback has a peak power output of 168 hp @ 5500 rpm, while the turbocharged version of the same car makes 250 peak hp @ 6000 rpm. If we convert these flywheel horsepower figures to the maximum energy output using equations in the previous section, we find that:

$$1 \text{ J/s} = 0.00134 \text{ hp}$$

$$168 \text{ hp} \times \frac{1 \text{ J/s}}{0.00134 \text{ hp}} \times \frac{1 \text{ min}}{5500 \text{ rev}} \times \frac{60 \text{ s}}{1 \text{ min}} \times \frac{2 \text{ rev}}{\text{cycle}} = 2.7 \text{ kJ/cycle}$$

$$250 \text{ hp} \times \frac{1 \text{ J/s}}{0.00134 \text{ hp}} \times \frac{1 \text{ min}}{6000 \text{ rev}} \times \frac{60 \text{ s}}{1 \text{ min}} \times \frac{2 \text{ rev}}{\text{cycle}} = 3.7 \text{ kJ/cycle}$$

These values are much lower than predicted by our ideal-gas model. However, our model calculations were performed assuming 100% conversion of heat into expansion work and are very close to these values if we also factor in the heat-engine efficiency rather than just the volumetric efficiency and bear in mind that a turbocharger is a device for improving that efficiency.

If we look more deeply at the factors that affect our earlier ideal-gas-based analysis of the effect of a turbocharger, we see that several of our key assumptions are likely invalid. For example, according to the ideal-gas model, if we assume that air has a constant molar volume (n/V) at all temperatures and pressures, then an increase in gas pressure must lead to an increase in the gas temperature. We know there will be some temperature increase in our inlet gas versus the normally aspirated case because the turbocharger will not be 100% efficient at converting the kinetic energy input by the impeller into potential energy, as some of this energy will go to heating the inlet gas. Furthermore, the device itself will heat the inlet gases simply due to indirect contact with the hot exhaust gases driving the impeller. Since the molar volume of the gases is in fact a function of temperature, the temperature increase of the intake air leads to a drop in gas density and a reduction in the moles of O_2 delivered versus our simple model. It is safe to say that this change in inlet gas temperature/density is a significant limitation preventing the theoretical energy gains per cycle predicted by our overly simplified model.

The problem of hot inlet gases can be partially addressed by installing an intercooler after the turbocharger. An intercooler is essentially a heat exchanger, a device that facilitates the cooling of some solid or fluid by efficiently dissipating heat to another solid or fluid. A complete discussion of heat exchange and heat exchangers is easily a topic for a full-semester course and is generally not covered in great detail in a typical chemistry curriculum. However, the qualitative function of an air-to-air heat exchanger, which is the most common type of vehicular intercooler, is relatively straightforward. Essentially, the intercooler seeks to maximize the temperature difference between the inlet gas and the metal of the air intake. This is because metals are good conductors of heat, and the rate at which the intake metal can remove thermal energy from the intake gas (and thus partially restore the initial gas density) is directly proportional to the temperature difference between the gas and the metal. If we place a large number of thin metal fins on the outside of the air intake, the fins will pull thermal energy from the part of the intake that is in direct contact with the hot inlet gases and spread it over a body with much greater surface area (Figure 2.8). This finned portion of the intake is typically placed in an area on the car where it will experience a large flow of cool external air perpendicular to the fins, i.e., in the front of the car or below a hood scoop. The flowing gases will reduce the temperature of the intake metal and maximize the rate at which our intake material can remove heat from the inlet gas, leading to increased gas density and thus greater engine power output. Even better, this design requires little to no extra energy input, since the cooling air will be moving relative to the vehicle anyway. The 250-hp version of the Subaru engine discussed earlier is also equipped with an air-to-air intercooler interfaced with a functional hood scoop, meaning that the true power gain due to the turbocharger is less than the 37% peak energy increase reported by Subaru. Clearly, reality is substantially more complicated than the simple chemistry models that we have applied in this section; however, simple chemical principles were able to teach us several lessons about the usefulness of forced-induction systems.

Figure 2.8 (See color insert.) Photo of the engine bay in a 2005 Subaru Outback 2.5 XT, highlighting the turbocharger (green oval) and the intercooler (red oval). The turbocharger is mostly hidden under a heat shield, but you can clearly see the turbo outlet bolted to the intake port of the intercooler (left side of the intercooler).

Water/Methanol Injection Systems

Another method for reducing the inlet gas temperatures in forced-induction systems is to use evaporative cooling of another fluid injected directly into the inlet gas stream. For example, injecting methanol or a water/methanol mixture into the inlet gas before or after the throttle body, or even directly into the combustion chamber, will cool the temperature of the inlet gases. The enthalpy of vaporization for water is 40.68 kJ/mol and the enthalpy of vaporization for methanol is 35.3 kJ/mol, meaning they can remove substantial heat during vaporization. These compounds also have moderate boiling points that allow them to be liquids at atmospheric temperatures, permitting you to hold a large number of molecules in a small space during storage, yet they easily vaporize at the inlet air temperatures of forced-induction vehicles (boiling point of methanol = 148°F, boiling point of water = 212°F). The evaporation rates are made very rapid in this type of cooling system by using an injection system that generates very tiny liquid droplets such that the liquid surface area is quite high, much like the fuel injectors in the engine. The cooler inlet gases provide a greater oxygen density to the combustion chambers, which permits one to burn more fuel in a given time, thereby increasing the power output of the engine. This evaporative cooling can also help prevent predetonation in forced-induction or high-compression engines, allowing users to optimize the engine timing for additional power output or to use lower octane fuels. Cooling the combustion gases will also reduce the temperature of the exhaust gases, which can help reduce the production of certain pollutants such as thermal NO_x (the NO_x produced when atmospheric nitrogen makes contact with very hot exhaust manifolds, catalytic converters, etc.). These injection systems can be used with or without intercoolers, and commercial systems are available for automobiles.

One additional take-home message from this exercise is that turbochargers (and to a lesser degree superchargers) can also be used to improve the fuel economy of vehicles. For example, let us say that you decide 150 hp is a suitable amount of power for a car and that your company makes a normally aspirated 2.0-L engine with that power output. In principle, you could also generate that 150 hp by using a smaller engine, say 1.5 L, and turbocharging it. Both engines would use the same amount of fuel under acceleration. However, simply based on displacement volume alone and our example from this chapter, the 1.5-L engine should consume less fuel at a constant cruising rpm than the 2.0-L version. However, the cruising rpm for a particular vehicular velocity is dependent upon the transmission and several other factors. It is more likely that the 1.5-L turbo will be turning higher rpm at the same cruise velocity as the 2.0-L engine, leading to a lower than predicted gain in fuel economy.

Nitrous Oxide Injection

Another way to increase engine efficiency by bypassing the normal aspiration mode of an engine is to directly inject an alternative oxidant into the combustion chamber. This is the key idea behind nitrous oxide injection systems for engines. Nitrous oxide (N_2O) improves fuel oxidation (allows more fuel to be burned in a given time interval) because of its stoichiometry. When 1 mole of nitrous oxide breaks down in a hot engine cylinder, you get 1 mole of N_2 (g) and ½ mole of O_2 (g), or a 33% oxygen atmosphere. This is significantly more oxygen to participate in combustion reactions versus the 21% found in a typical atmospheric volume. In addition to the higher saturation of O_2, the evaporation of liquid N_2O in the intake system provides substantial cooling of the inlet gases, increasing their density and the volumetric efficiency of the engine. The downsides to nitrous oxide induction are that nitrous systems only operate when you have nitrous oxide in the onboard tanks (as opposed to the continuous operation of turbo and superchargers), tank pressure fluctuations require some effort to control, and higher stresses are put on engine components.

2.7 Biodiesel: Turning waste into energy

Chemistry Concepts: organic chemistry, viscosity, thermochemistry
Expected Learning Outcomes:
- Describe the chemical process of producing biodiesel
- List the challenges of running a diesel engine on pure vegetable oil
- State the pros and cons of fueling a vehicle with plant oils or plant-oil-derived biodiesel

This chapter focuses on combustion and the fossil fuel–based engines. Thus, a complete discussion of alternative fuels such as bioalcohols, methane, hydrogen, etc., is left to the many environmental chemistry, green chemistry, and related texts that already cover these topics in some detail.

However, in keeping with our discussion of improving IC engine efficiency and power, we note that diesel engines are more efficient as delivered and that the overall benefits of diesel engines may be further improved by converting consumer waste streams into fuel sources. This section of the chapter examines alternative fuels for diesel engines and the chemistry of biodiesel production.

Diesel fuels have much larger molecules on average than gasoline, and some of them start to approach the size of fatty acid molecules found in vegetable oils (10–20 C atoms per chain). Not surprisingly given this similarity, diesel engines can run on either pure vegetable oils or chemically modified vegetable oils. The biggest obstacle to using straight vegetable oil in a diesel engine is the ability of the oil to flow (viscosity). Many types of vegetable oil tend to have gel points—the temperature at which the oil turns into a thick, gelatinous solid—that are above the cold operating temperatures of engines in some winter climates. When it gets cold and the oil gels, it will obviously not flow and cannot be delivered to the engine in a state where it is ready to combust. To get around this aspect of using pure vegetable oil, many straight vegetable oil diesels have oil preheaters. As you can observe in your frying pan, a plant oil that has been sufficiently warmed flows easily like water or gasoline. It is also important to filter the vegetable oil and remove any particulates or contaminants that may clog the fuel delivery system or contribute to deposit buildup in the engine and exhaust systems. Many vegetable oil diesels run two fuel tanks, one for the vegetable oil and one that contains pure petroleum-derived diesel fuel to run during engine startup and just before engine shutdown, which helps to mitigate some of the plant oil viscosity and carbon deposit issues associated with pure plant oil as a fuel. Pure vegetable oils also have other downsides, such as a greater reactivity with atmospheric oxygen than petroleum-derived fuels, leading to low stability and a propensity to contaminate the lubricating oils in the engine. Some of these drawbacks can be offset by mixing vegetable oil and conventional diesel fuels, though one must ensure the physical properties of the mixture remain suitable for use as a fuel. Most blends must remain at or below 30% vegetable oil to function properly. Details about engine modifications and other design criteria for converting a diesel engine to an efficient power plant running on pure plant oils are discussed in other texts.

The other possibility for using new or waste vegetable oil is to convert the oil into biodiesel, a chemical process that produces a fuel quite similar to petroleum diesel in many ways. Most vegetable oils contain primarily biological polymers with relatively short carbon chains (10–25 C atoms), called *fatty acids*, attached to a short chemical backbone called *glycerin* or *glycerol*. These triglycerides are the critical reactant in the vegetable oil that will be converted into fuel. Plant oils often contain some degree of free fatty acids as well, or acids that exist as single chains that bear an intact carboxylic acid functional group. If one is using waste vegetable oil to make biodiesel, the first step is to filter and perhaps use other pre-processing cleaning methods to remove bits of food and other solids from the oil. Typically, preprocessing involves physical approaches rather than chemical approaches and thus will not be discussed further here. After the oil is clean, a small sample can be taken, warmed, and titrated with a strong base such as sodium hydroxide to determine the free fatty acid content. The next step is to conduct a chemical process called a *transesterification reaction* that will separate the fatty acid molecules from the triglycerides, yielding fatty acid esters and glycerol:

This is typically done by mixing an alcohol into the oil. Methanol is the most common alcohol used in the preparation of biodiesel, as it is the most readily abundant and cheap alcohol, though any short-chain alcohol will perform the transesterification reaction. This type of transesterification is often a very slow process because of the resistance to deprotonation of the alcohol, and thus a strong base is often added to the mixture. The base serves two functions, the first of which is neutralizing the free fatty acids and the second of which is to deprotonate the alcohol to make a reactive alkoxide anion. The mixture of plant oil, alcohol, and base is mixed and allowed to react. The result is two separate organic layers, one being the esters of the fatty acid chains and the other being a heavier layer of glycerin contaminated with impurities. The fatty acid ester layer is removed from the system and is essentially ready to use. In some biodiesel production strategies, this fatty acid ester layer is washed with water, which removes any water-soluble impurities from the mixture, including residual base and alcohol. The alcohol can also be recovered by distilling the unwashed fatty acid ester layer. Once the water layer and fatty acid esters have separated, the fatty acid esters are dried and stored for use in a diesel vehicle.

Biodiesel offers several advantages and disadvantages when compared to conventional petroleum-derived diesel fuel. One relates to the amount of sulfur in the fuel. Biodiesel is essentially a sulfur-free fuel, while diesel fuels come in several grades that all bear some sulfur. The lack of sulfur in biodiesel means that it does not contribute significantly to SO_x gases that stimulate acid rain. In addition, studies have shown that pure biodiesel generates lower levels of polyaromatic hydrocarbons (PAHs) that are thought to be carcinogens, low unburned hydrocarbons, low CO, low CO_2, and low NO_x. However, biodiesel contains different types of molecules than petroleum-derived diesel and therefore may accelerate the dissolution of organic materials, including rubber hoses, rubber fuel lines, gaskets, etc. Some of the benefits of biodiesel can also be realized by fueling a vehicle with biodiesel and petroleum diesel blends. Several standard categories of mixture exist, and B-20 (a 20% biodiesel mixture) has also been shown to outperform petroleum diesel in terms of reduced pollutant gas emissions. As is the case for ethanol discussed earlier, whether biodiesel is truly an energy benefit requires a complete life-cycle analysis. However, converting a waste stream such as used vegetable oil into a fuel quite likely turns out to be an energy advantage and is a great project to start on college campuses.

chapter three

Oxidation and reduction

Cars are subject to (or take advantage of) oxidation/reduction (REDOX) reactions—reactions where electrons are transferred between chemical species—every day. Several examples of important REDOX reactions in cars include the combustion process in fossil fuel-burning engines, the electrochemistry involved in batteries and fuel cells, and the corrosion of metal car components. Because REDOX reactions always involve a transfer of electrons, REDOX chemistry may be used to generate a current, as in a battery, or be a relatively unnoticed aspect of a chemical reaction, as it often is in combustion or corrosion processes. REDOX chemistry is also essential to metal plating and corrosion protection in vehicles. In this chapter, we will focus on understanding the details of these automotive applications of REDOX chemistry.

3.1   A second look at combustion

Chemistry Concepts: oxidation numbers, REDOX terminology, activity series

Expected Learning Outcomes:
- Explain basic REDOX terminology
- Identify what is being oxidized and reduced in a combustion reaction

In every REDOX process, some element or chemical component gains electrons from another element or chemical component in the system. The species that loses electrons is said to be *oxidized* and the species that gains electrons *reduced*, and a REDOX reaction always involves both oxidation and reduction. The origin of these terms relates to the chemical concept of oxidation numbers, which is a numerical system to track the flow of electrons in a chemical reaction, assuming that any electron transfer is complete. In other words, partial charges are not allowed. In most cases, the oxidation number of an ion is identical to its ideal charge in a particular substance, though the transition metals and many nonmetals often have several possible oxidation states at most temperatures and pressures. Rules for assigning oxidation numbers in chemical reactions can be found in general chemistry textbooks and are thus not discussed in detail here. When a substance is reduced, its oxidation number decreases.

That does not necessarily mean the oxidation number becomes negative; it can be a smaller positive number. Since we cannot change the number of positive charges in an atom without changing the chemical identity of the substance, we can only decrease an oxidation number by adding electrons to that chemical component. When a substance is oxidized, the oxidation number increases (becomes more positive), meaning that it has given electrons away to some other substance. If you can remember that reduction means the oxidation number gets more negative via the gain of electrons, you should have no trouble remembering that oxidation is the opposite process where the oxidation number increases.

Chemists also talk about REDOX reactions based upon the role substances play in a reaction. For example, a substance that is easily oxidized (gives up electrons easily) is a good reducing agent, that is, it is effective at making the oxidation number of other species decrease. Likewise, a substance that happily accepts electrons is a good oxidizing agent, that is, it facilitates other elements giving up electrons by providing a stable place for the electrons to go. Often, we arrange substances by the ease with which they are reduced in a table known as an electrochemical series or activity series (Table 3.1). Simple inspection of such a table can tell you whether a REDOX reaction will occur between a given set of chemicals at

Table 3.1 A Typical REDOX Activity Series

Easily Oxidized
cesium
rubidium
potassium
sodium
calcium
magnesium
aluminum
titanium
manganese
zinc
chromium
iron
nickel
lead
copper
gold
Easily Reduced

Note: Elements are arranged on a continuum ranging from easily oxidized (top) to easily reduced (bottom).

a given set of conditions. A good rule of thumb for using these tables is to find the precious metals, those that are often used in jewelry and metal coins. Precious metals are all easily reduced to their metallic state, otherwise gold and platinum wedding bands would likely dissolve in the first rainstorm. Precious metals are good oxidizing agents and thus they prefer to accept electrons from other species and transform from ions in solution to a pure metallic state when electron transfer is possible. Species on the opposite end of the activity series are easily oxidized, making them good reducing agents. These chemical species prefer to go from a solid state to a solvated ion when reacted with materials on the activity series that are closer to the precious metals.

When writing REDOX reactions, often we use both the overall reaction and two half-reactions, one associated with the oxidation process and the other with the reduction process. The reduction and oxidation half-reactions allow us to view which species are generating electrons and accepting electrons and see the number of electrons that are transferred. They are called half-reactions because, if they are added together, they generate the overall reaction for a process. For example, if we examine the reaction of zinc metal and a copper(II) sulfate solution, the overall reactions is:

$$Zn\ (s) + CuSO_4\ (aq) \rightarrow Cu\ (s) + ZnSO_4\ (aq)$$

In this reaction, the zinc metal is oxidized and the Cu^{2+} ion in solution is reduced. We can show these two steps and the associated electron transfers by writing a half-reaction associated with each part of the REDOX process:

$$Zn\ (s) \rightarrow Zn^{2+} + 2e^-$$

$$Cu^{2+} + 2e^- \rightarrow Cu\ (s)$$

The zinc reaction is called the oxidation half-reaction, since it describes a release of electrons (electrons are on the product side). The copper reaction is the reduction half-reaction, since the oxidation number of Cu^{2+} is reduced as it gains electrons (electrons are on the reactant side). The sulfate doesn't appear in either of these half-reactions as written because it is not donating or receiving electrons, though we could certainly write the charged ions as aqueous sulfate compounds. It is assumed that you understand how to assign oxidation numbers and understand the basic REDOX terminology of these first two paragraphs in the remainder of this chapter.

As suggested in the title of this section, the combustion reaction that we discussed in the previous chapter is also an example of a REDOX reaction. Let's take another look at the combustion of hydrocarbons discussed earlier using the REDOX definitions and terms discussed thus far. The overall chemical reaction for the combustion of the simplest hydrocarbon, methane, is:

$$CH_4 + 2O_2 \rightarrow CO_2 + 2H_2O$$

If we look at the oxidation numbers of each element on the reactant and product side of this reaction, we can see if there is any electron transfer occurring and what substances play a role in the electron transfer. The oxidation number of hydrogen in octane can only be +1 or −1, and in this case it must be +1 according to the accepted oxidation number rules. This means that the oxidation number of carbon on the reactant side of this reaction is −4, because methane must be a neutral compound and there are four total positive charges, one due to each of the four positively charged hydrogen atoms in the structure. The oxygen on the reactant side has an oxidation number of zero because O_2 (g) is the reference state of oxygen at normal atmospheric temperature and pressure. By definition, the *reference state* for each element under a given set of conditions is the most stable state of the element at the pressure and temperature of the reaction. Reference states always have oxidation numbers of zero. On the product side, we see that the carbon in CO_2 has an oxidation number of +4, since the two oxygen atoms in CO_2 have −2 oxidation numbers and the overall molecule must remain neutral. Similarly, in H_2O the oxygen has a −2 oxidation number, making the two hydrogen atoms each have a +1 oxidation number. Thus, the combustion of methane is a REDOX process involving the transfer of eight electrons from carbon to oxygen atoms.

Now let's look at octane combustion again. The overall reaction is:

$$C_8H_{18} + {}^{25}/_2O_2 \rightarrow 8CO_2 + 9H_2O$$

The oxidation number of hydrogen in octane is +1, just as in the case for methane. Thus, octane has a total positive charge of +18 due to the 18 hydrogen atoms in the molecule. Since octane is a neutral substance, this means that the eight carbon atoms must account for 18 negative charges, or that each carbon atom on average has a −2.25 charge. However, by rule, we always set oxidation numbers to be integers, since they are calculated assuming complete electron transfer. It is often difficult to determine the exact oxidation number of any one carbon in an organic compound, so typically the oxidation state of carbon over the entire compound is reported instead. For octane, this means we would say carbon in octane has an oxidation number of −18. To illustrate this point, if you draw the structure of *n*-octane (Figure 3.1), you will see that the two carbons at the ends of the chain that have bonds to three hydrogen atoms have oxidation numbers of −3 and that the six intermediate carbons have charges/oxidation numbers of −2 since they are bonded to only two hydrogen atoms. Carbon in CO_2 on the product side has an oxidation number of +4, as it did for methane. The oxygen on the reactant side still has an oxidation number of 0, with all oxygen atoms having an oxidation number of −2 on the product side. The oxidation number of hydrogen is unchanged in the reaction. Thus, in our combustion reaction, the central and edge carbon atoms in

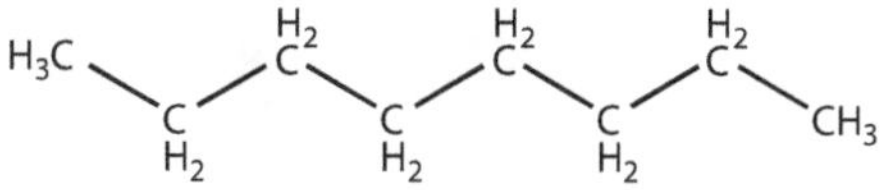

Figure 3.1 Structure of *n*-octane.

octane are oxidized into eight CO_2 molecules (total C oxidation number of $8 \times +4 = +32$), meaning there was a total electron donation of 50 e^- from the C in octane during combustion. The oxygen from the oxygen gas accepts all of these electrons as it is reduced from a zero oxidation number to a –2 during the reaction. Thus, in addition to being a highly exothermic reaction capable of doing a significant amount of expansion work, octane combustion is a REDOX process involving electron transfers between carbon and oxygen.

3.2   Batteries

Chemistry Concepts: electrochemistry, REDOX reactions, solution chemistry, standard electrode potentials

Expected Learning Outcomes:

- Explain the function and purpose of an electrochemical cell
- Explain the lead-acid battery and how it works
- Mention some more recent battery technologies and their basic chemistry

Batteries are devices that can store and release energy via electrical currents and REDOX reactions. Essentially, they take advantage of electrochemistry to convert energy between chemical and electrical forms. There are many different types of batteries, and the battery type best suited to a particular application depends on the amount of work required, the ease and speed with which the battery can be recharged, the size and weight of the cell, and whether or not the battery itself needs maintenance to function optimally or properly. The two most common types of batteries found in cars are the lead-acid batteries of conventional combustion-based vehicles and the advanced Li-ion batteries found in electric cars like those produced by Tesla Motor Company. In both cases, the batteries are electrochemical cells that generate a flow of electrons due to REDOX reactions.

To understand electrochemical cells, let us talk about the most common type of commercial battery, the alkaline battery (Figure 3.2). In a typical alkaline battery, zinc metal and manganese(IV) oxide are brought into contact via an alkaline solution of potassium hydroxide. The zinc metal and manganese(IV) oxide are known as the electrodes and are the locations where the REDOX reactions take place. The potassium hydroxide

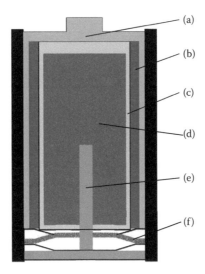

Figure 3.2 Cross section of a typical alkaline battery. (a) is the current pickup for the cathode, (b) is the MnO_2 cathode material, (c) is a separator membrane that is permeable to the K^+ and OH^- ions, (d) is the Zn metal anode, (e) is the anode current pickup, and (f) is a separator/cap that permits expansion and keeps a good seal. The electrolyte soaks the cathode and anode, which are both granulated or powdered materials.

solution is called an electrolyte, or a solution bearing charged species as ions, that is capable of transporting charge. Potassium hydroxide makes an effective electrolyte in water because potassium hydroxide is a strong base that completely dissociates into K^+ and OH^- ions in this solvent. Once the electrodes are connected via a wire (or in the case of an alkaline battery, a wire connects the two terminals), electrons generated by REDOX reactions are allowed to flow between the electrodes. Electrons flow from the anode, which is the electrode where oxidation takes place and electrons are generated, to the cathode, which is the electrode where the reduction reaction takes place. In our alkaline cell, the zinc is the anode and the manganese(IV) oxide is the cathode:

$$Zn\ (s) + 2OH^- \rightarrow ZnO\ (s) + H_2O\ (l) + 2e^- \qquad E^0 = -1.28\ V$$

$$2MnO_2\ (s) + H_2O\ (l) + 2e^- \rightarrow Mn_2O_3\ (s) + 2OH^- \qquad E^0 = +0.15\ V$$

The zinc metal anode gives up two electrons when it reacts with the hydroxide ions to form the zinc oxide, and the manganese metal accepts one electron as it converts from manganese(IV) oxide to manganese(III) oxide at the cathode. As the ions of the electrolyte diffuse toward each electrode, the electrolyte plays a critical role in preventing a buildup of electrons that would eventually generate a null current. The negative ions flow toward the anode (the hydroxide toward the zinc in our example) as the anode is depleted of negative charge due to the flow of electrons, and the positive ions diffuse toward the cathode to balance the negative charge that develops as it receives the electrons. Thus, for any electrochemical cell to generate a current, we need a cathode where reduction will take place, an anode where oxidation will take place, a wire connecting the electrodes that permits electrons to flow between them, and contact of the electrodes via an electrolyte where ion diffusion helps to prevent charge buildup (Figure 3.2). Typically, the electrodes are either in the same electrolyte solution, or two different electrolyte solutions are connected via a salt bridge, a device that permits a flow of ions without allowing the different electrolytes to mix. We consider each electrode and its associated reactions to be a half-cell, since two such half-cells are required to produce a single functional electrochemical cell. Remember, we need an oxidation step and a reduction step!

Each half-reaction in our alkaline battery example is associated with one of the half-cells and is followed by a voltage. These voltages are the standard half-cell potentials (denoted with the variable E^0) and are determined by measuring the cell potential when a half-cell is paired with the standard hydrogen electrode (SHE) to generate an electrochemical cell. A full discussion of the SHE is often found in analytical or general chemistry texts, and for our discussion it will suffice to know that the SHE is a reference electrode and assigned a standard cell potential of zero at all temperatures. Voltages can only be measured for complete electrochemical cells, and by making a cell using the SHE electrode, the voltage for the resultant electrochemical cell can be completely assigned to the non-SHE half-cell. For many half-cells, these standard cell potentials are tabulated and can be used to estimate the total electromotive force (EMF) that will be generated by pairing a given set of half-cells (the standard cell potential, E^0_{cell}) using the following equation:

$$E^0_{cell} = E^0_{cathode} - E^0_{anode}$$

Thus, our alkaline battery should generate a voltage equivalent to 0.15 V − (−1.28 V), or 1.43 V. If you have a voltmeter and connect the probes to either side of an alkaline battery, this should be very close to the voltage that you measure. Table 3.2 includes a series of important REDOX reactions in automotive chemistry and their associated reduction potentials.

Knowing the cell EMF is useful, but how much energy do you get from an electrochemical cell? Does the size of a cell relate to its voltage? And how do you know if the cell produces this energy spontaneously, as required for a useful battery? The answers to these questions can be found by exploiting the close relationship between the thermodynamics

Understanding chemistry through cars

Table 3.2 A Table of Select REDOX Potentials with Applications in Automotive Chemistry

Reaction	E^0 (V) @ 25°C
Reduction Least Likely	
$Li_nC \rightarrow Li_{(n-1)}C + Li^+ + e^-$	−3.04
$Al(OH)_3 + 3e^- \rightarrow Al + 3OH^-$	−2.31
$Al^{3+} + 3e^- \rightarrow Al$	−1.662
$2H_2O + 2e^- \rightarrow H_2 + OH^-$	−0.8277
$Cr^{3+} + 3e^- \rightarrow Cr$	−0.744
$Fe(OH)_3 + e^- \rightarrow Fe(OH)_2 + OH^-$	−0.56
$ZnOH^+ + H^+ + 2e^- \rightarrow Zn + H_2O$	−0.479
$Fe^{2+} + 2e^- \rightarrow Fe$	−0.447
$Pb + HSO_4^- \rightarrow PbSO_4 + H^+ + 2e^-$	−0.36
$PbSO_4 + 2e^- \rightarrow Pb + SO_4^{2-}$	−0.3588
$CrO_4^{2-} + 4H_2O + 3e^- \rightarrow Cr(OH)_3 + 5OH^-$	−0.13
$Fe^{3+} + 3e^- \rightarrow Fe$	−0.037
$2H^+ + 2e^- \rightarrow H_2$	0
$CoO_2 + Li^+ + e^- \rightarrow LiCoO_2$	0.36
$Fe(s) + 2e^- \rightarrow Fe^{2+}$	0.41
$Cr_2O_7^{2-} + 14H^+ + 9e^- \rightarrow 2Cr(s) + 7H_2O(1)$	0.59
$Zn + 2e^- \rightarrow Zn^{2+}$	0.76
$Fe^{3+} + e^- \rightarrow Fe^{2+}$	0.771
$Pt^{2+} + 2e^- \rightarrow Pt$	1.18
$O_2 + 4H^+ + 4e^- \rightarrow 2H_2O$	1.229
$Cr_2O_7^{2-} + 14H^+ + 6e^- \rightarrow 2Cr^{3+} + 7H_2O$	1.33
$HCrO_4^- + 7H^+ + 3e^- \rightarrow Cr^{3+} + 4H_2O$	1.350
$PbO_2 + SO_4^{2-} + 4H^+ + 2e^- \rightarrow PbSO_4 + 2H_2O$	1.6913
$PtO_3 + 2H^+ + 2e^- \rightarrow PtO_2 + H_2O$	1.7
Reduction Most Likely	

Note: Reactions are arranged on a continuum ranging from the least likely reductions (top) to the most likely reductions (bottom). Values are drawn from D. R. Lide, ed., *CRC Handbook of Chemistry and Physics*, 80th ed. (Boca Raton, FL: CRC Press, 1999).

of the chemical system and the system electrochemistry. In thermodynamics, the Gibbs energy of a reaction is a state function that describes the maximum amount of nonexpansion work that a certain reaction can perform under constant pressure conditions. It is related to a state function called *enthalpy* that describes the heat evolved under constant pressure and a second state function called *entropy* that describes the spread of energy over available energy states of molecules in the reaction system. Together, the enthalpy change of the system and the entropy change of the system define the total entropy change of the universe during a chemical reaction. Because all processes proceed toward maximum entropy spontaneously, the Gibbs energy can also be used to determine whether a reaction proceeds spontaneously. Negative values of the Gibbs energy are associated with spontaneous chemical reactions. Because electrical work is not expansion work, the Gibbs energy for the REDOX reaction in an electrochemical cell tells us the energy produced by that cell that is free to do electrical work. From the laws of thermodynamics, one can derive a relatively easy formula for converting between the Gibbs energy of a reaction and the voltage that a cell can produce:

$$\Delta G^0_{cell} = -nFE^0_{cell}$$

where ΔG^0_{cell} is the standard Gibbs energy change, n is the number of electrons transferred per mole of products, F is Faraday's constant, and E^0_{cell} is the standard cell potential. This means that you can determine the voltage that will be generated in a battery simply by finding the standard entropies and enthalpies of formation for each compound (or the Gibbs energies of formation for each compound) in thermodynamic tables, adjusting the values to the operating temperature of the battery, and using concepts we discussed in Chapter 2 to determine the Gibbs energy change.

$$\Delta G^0_{rxn} = \Delta H^0_{rxn} - T\Delta S^0_{rxn} = \sum_{products} n\Delta G^0_f - \sum_{reactants} n\Delta G^0_f$$

Likewise, you can use measurements of potential to determine the free energy of the reaction involved in an electrochemical cell. You can also use your cell voltage to determine if you have a spontaneous electrochemical cell suitable for a battery, called a *galvanic cell*. Since spontaneous processes always generate negative Gibbs energies of reaction, inspection of the Gibbs energy/cell-voltage relationship shows that spontaneous reactions will always generate positive standard cell potentials, as in the case of our standard alkaline battery. The thermodynamics also show that the voltage generated by a cell has nothing to do with the size of the cell itself, just the reaction stoichiometry. You can prove this by comparing alkaline batteries of various sizes. Fresh AA and AAA alkaline batteries will generate the same voltage on a voltmeter. The size of the battery instead relates to the amounts of reactants and therefore how long the battery will generate that specific voltage. Although you can always use tables of standard cell potentials to determine the voltage a battery will generate, one should always remember that the energy released is the result of a chemical reaction and thus relatable to thermodynamic principles.

Let's now look at the most common battery found on the road today, the lead-acid battery. Car batteries must be capable of providing a very high current required to turn the starter motor (hundreds of A), must be rechargeable, and must generate a voltage compatible with the vehicle electronics (typically 12 V). Six-cell lead-acid batteries are ideally suited for this application. In a single-cell lead-acid battery, plates of lead oxide and lead are exposed to a single concentrated sulfuric acid electrolyte solution, typically ≈ 3 M H_2SO_4. The lead plate serves as the anode and the lead oxide plate as the cathode, with both plates converting to lead sulfate during discharge.

$$Pb(s) + HSO_4^- \rightarrow PbSO_4(s) + H^+ + 2e^- \qquad E^0 = -0.36 \text{ V}$$

$$PbO_2(s) + HSO_4^- + 3H^+ + 2e^- \rightarrow PbSO_4(s) + 2H_2O(l) \qquad E^0 = 1.69 \text{ V}$$

Applying what we know about electrochemistry, we see that this cell spontaneously generates an EMF of ≈ 2 V. One can also see that the sulfuric acid is consumed during discharge and leaves behind an electrolyte severely depleted in acid protons and sulfate anions. Another discharge effect less easy to see in the reaction stoichiometry is that the plates grow in size during discharge, and they may in fact contact one another if elongated, fingerlike $PbSO_4$ growths called *dendrites* extend and cause two plates to touch. For this reason, lead acid batteries also include porous separators that prevent plate-to-plate contact while permitting electrolyte to touch both plates. Since each individual cell only generates 2 V and the overall car battery needs to generate 12 V, we can obtain the voltage we need by wiring six separate cells together in series (see a basic physics text for information on circuits and their properties). When the battery recharges, an external source of potential is used to forcibly drive the reactions in the opposite direction, regenerating the acid electrolyte and converting the plates back to lead and lead oxide.

Chemical Struggles for Your Battery in Winter

Anyone who drives in climates where the weather can drop below freezing knows that car batteries struggle to start the engine when it is cold outside. If you measure the voltage on your battery in winter, you might get a reading that is normal ($\approx 12.6 \pm 0.2$ V) despite the difficulty your battery has in turning over the engine. Why is this? The answer lies in the fact that starting your engine relates more directly to the power that your battery can deliver.

For the car to start, the battery must deliver enough electrical power to provide the starter motor the torque it needs to rotate the crankshaft until the engine fires. Battery companies give you some estimate of the wintertime power a battery can generate using a parameter called the *cold cranking amps* (CCA), which is the current that your 12-V battery can generate when the outdoor temperature is 0°F or –18°C. Recall that electrical power is the product of the current and the voltage. The current that the battery can generate on startup is related to its internal resistance, or really the ease with which the electrolyte can transport charge between plates within the battery itself. We call this property the *conductivity* of the battery. There are several chemical reasons why the conductivity within your battery drops when it is cold. The first relates to development of concentration gradients, or spatially dependent concentrations, in the sulfuric acid electrolyte. In the winter, the heater, defroster, wipers, and other hungry electrical accessories can put a pretty significant load on the battery. If you drive mostly on short trips, the car may not have enough time to fully recharge the battery, leaving part of the plates rich in $PbSO_4$ and the nearby electrolyte depleted in acid. In addition, concentrated sulfuric acid has a density nearly twice that of water thanks to the heavy sulfur atom, the many oxygens in sulfate, and the ease with which sulfate can participate in hydrogen bonds, meaning that the regenerated acid during recharging migrates to the bottom of the battery. Over time in winter, the lack of full recharge and the increased viscosity (resistance of a fluid to flow) of the electrolyte solution at cold temperatures may produce a concentration gradient in the sulfuric acid. In regions with low acid, the conductivity is lower because there are fewer ions that can transport the charge between plates, leading to a drop in battery conductivity and the CCA. The change in surface area experiencing good electrical contact does not affect the cell voltage, however, since the cell voltage does not depend on the quantity of material, just the chemicals involved, so your voltage measurement appears within proper specifications. In principle, inverting the battery several times may remove the sulfuric acid gradient and restore some degree of CCA to the battery in the winter, but this action does not occur during normal operation of a vehicle.

A second chemical concern is the formation of crystalline $PbSO_4$ on the cell plates. Your battery typically produces a finely divided amorphous $PbSO_4$ phase on the surface of the cell plates. Unfortunately, any portion of the plate surface subject to high acid regions during winter can cause this amorphous $PbSO_4$ phase to convert to crystalline material.

This well-ordered $PbSO_4$ is less conductive than the Pb/PbO_2 and much more difficult to convert back to Pb or PbO_2 during recharge, making plate conversion to $PbSO_4$ in the presence of very strong sulfuric acid effectively irreversible. Another chemical problem for your battery in winter is that the low-acid regions are subject to their own set of corrosion reactions that do not occur when the acid concentration is relatively uniform. For example, conversion of the Pb plate to PbO or PbO_2 by direct reaction with oxygen will lead to local increases in resistance within the cell.

The subsequent drop in conductivity due to these three unwanted chemical processes lead to reductions in a battery's CCA. The corrosion/sulfation reactions are generally not a problem when the battery is warm in the summer, as the sulfuric acid and water viscosities are reduced, leading to more rapid diffusional mixing of the acid electrolyte and uniform electrolyte conductivity. Over time, the chemical damage to the plates as a result of the cold weather acid gradient accumulates, eventually leading to a large enough reduction in CCA that the battery simply cannot provide the power needed to start the car in winter despite producing the appropriate voltage.

Though lead-acid batteries are common in vehicles and are very well suited to that role, they have some drawbacks. The lead used in these materials represents an environmental hazard, particularly if the battery is disposed of improperly at the end of its usable life. The plates can also crack over time due to repeated discharge–recharge cycles, limiting the output and life of the battery (see the sidebar on winter batteries for other chemical processes that reduce battery life). Lead-acid batteries are also very heavy due to the high density of lead, and are thus not good candidates for fully electric car batteries. The weight of an electric vehicle carrying an appropriate number of lead-acid batteries to generate the required current and voltage to drive a reasonable distance would make owning such a vehicle prohibitive. It is also possible for a lead-acid battery to explode if it is overcharged. Overcharging can generate hydrogen gas in the battery chamber through another electrochemical process, and this gas could ignite upon engine startup.

To overcome many of these challenges, fully electric cars currently make use of a more recently developed battery technology known as *lithium-ion batteries*. Lithium-ion batteries can be designed to be rechargeable, can be recharged quickly, continue to generate their initial voltage after many charge–discharge cycles, and offer a high energy density (a large amount of energy in a small, light package). They also offer relatively high efficiencies, with the industry target being 90% recovery of energy stored in the battery, which when combined with electric motors of nearly 80% efficiency give electric cars that can, in principle, operate at 72% efficiency (recall that gasoline IC engines are typically ≈20% efficient). Rechargeable Li-ion batteries nearly all have electrodes made of sheetlike materials that store lithium between layers and are known as intercalation compounds because the lithium ions are intercalated between the sheets. As the Li-ion battery is discharged, lithium ions flow out from between the layers of the cathode intercalation compound to the anode, and as the battery is recharged, they flow back into the interlayer spaces of the cathode. Movement of the lithium ions is facilitated by an organic electrolyte that can easily form complexes with lithium ions. There are many different types of anode/cathode materials that can be used for the Li-ion battery, and there are even more possibilities if we vary the electrolyte chemistry. The remainder of our discussion here will focus on the most widely used commercial formula at the writing of this text, the lithium–cobalt oxide battery. The half-reactions in the discharge process are:

$$Li_nC \ (s) \rightarrow Li_{(n-1)}C + Li^+ + e^- \qquad E^0 = -3.04 \ V$$

$$CoO_2 \ (s) + Li^+ + e^- \rightarrow LiCoO_2 \ (s) \qquad E^0 = +0.36 \ V$$

and show that this type of Li-ion battery generates a voltage of ≈3.4 V/cell. This is much higher than a single cell of an alkaline battery or a lead-acid battery. Application of the Gibbs energy/voltage relationship shows us that these batteries also provide a large amount of energy available for electrical work:

$$\Delta G^0_{cell} = -nFE^0_{cell} = -1 \times 96,500 \frac{J}{V} \times 3.4 \ V = 328 \ kJ$$

Given the fact that the Li-ion batteries are also much lighter than lead-acid batteries, we can store a substantial amount of potential electrical energy in a car with lithium-ion batteries and make it readily available by wiring large numbers of these cells in series and parallel. For example, according to the Tesla Motor Company literature from 2013, the Tesla battery pack in the 85-kW·h model S contains ≈7000 Li-ion battery cells that produce 3.6 V each, wired in both series and parallel fashions to produce a total pack voltage ≈375 V and store an amount of energy equivalent to 306 MJ. The battery pack is also warranted for 8 years, which suggests the confidence of the battery manufacturer (Panasonic) in the discharge–recharge capabilities of the Li-ion battery.[*]

Hydrogen Fuel Cells

In the last chapter, we discussed engine efficiency and how internal combustion engines are fairly inefficient power sources. Electric vehicles seem like a major step up, but the effective efficiency depends significantly on where an electric car gets its electricity. For example, the 72% efficient electric car we discussed earlier charged by a 40% efficient coal-burning power plant through a 90% efficient AC-DC converter has a net efficiency of only 26% from fuel source to end use, while the same car charged at a DC solar charging station may exhibit significantly higher efficiency. Another alternative to grid or solar charging is to run an electric vehicle powered by an onboard fuel cell, a device that generates electric current via REDOX reactions of reactants that are constantly replenished. Good fuel cell reactant materials are abundant, such as H_2 and O_2 reacting to produce water, and produce no harmful by-products. Alkaline hydrogen/oxygen fuel cells (fuel cells that react hydrogen gas and oxygen gas with an electrolyte solution containing OH^-) have long been used in space flight. However, the leading fuel cell candidate for automotive applications is the polymer electrolyte membrane (PEM) hydrogen fuel cell, which operates at lower temperatures and eliminates the need for the caustic electrolyte solution. It has been shown that a PEM hydrogen fuel cell–powered vehicle can achieve up to 40% efficiency. The PEM fuel cell generates electricity by reacting hydrogen gas and oxygen gas that have been catalytically split at the electrodes, generating water as a by-product:

$$\text{Anode: } H_2 \text{ (g)} \rightarrow 2H^+ + 2e^- \qquad E^0 = 0.00 \text{ V}$$

$$\text{Cathode: } \tfrac{1}{2}O_2 \text{ (g)} + 2H^+ + 2e^- \rightarrow H_2O \qquad E^0 = 1.229 \text{ V}$$

The porous polymer barrier must allow a flow of protons but not permit a flow of electrons, which travel via wires and do the work of turning the electric motors. Effective catalysts for splitting the hydrogen and oxygen gases are also required, and at this time platinum seems to be the best option for both

---

[*] Panasonic Corporation, "Lithium-Ion Batteries Technical Handbook," 2007, http://industrial.panasonic.com/www-data/pdf/ACI4000/ACI4000PE5.pdf (accessed June 1, 2014).

gas-splitting reactions. In principle, there are no harmful by-products and no direct CO_2 footprint (following manufacture and delivery) for an electric vehicle powered by a fuel cell. However, safety issues related to the possible explosion of a compressed hydrogen gas cylinder during a vehicular accident and the need to develop a brand new and widely available H_2 infrastructure have been major drawbacks to this approach. Alternative fuel cell chemistries are being explored to overcome these limitations (methanol, ethanol, etc.), but these are not ready for production vehicles at this time. It appears that limited quantities of hydrogen-based fuel cell vehicles are ready to come online despite these concerns (see news about the 2015 Hyundai Tucson fuel cell vehicle http://www.caranddriver.com/news/2015-hyundai-tucson-fuel-cell-photos-and-info-news), but by some estimates we are still 10–15 years from seeing regular operation of fuel cell vehicles on the highway.

3.3   *The catalytic converter*

Chemistry Concepts: gas laws, kinetic molecular theory, heterogeneous catalysis, REDOX reactions

Expected Learning Outcomes:

- Describe the primary goal of a catalytic converter
- Write common chemical reactions that occur in a catalytic converter
- Explain the parts of a catalytic converter
- Explain when cars generate the most pollution and how that relates to the catalytic converter

Another place where REDOX chemistry plays a critical role in a car is the catalytic converter. A catalyst is a substance that increases the rate of a chemical reaction without itself being consumed. Functionally, the catalytic converter in your car increases the rate of many chemical reactions that convert harmful exhaust gases to more inert gases and CO_2. Here, *harmful gases* refer to combustion by-products that are toxic or otherwise directly harmful to people and the environment as well as those that play key roles in the chemistry of tropospheric smog or acid rain. Examples of such gases include volatile organic compounds (VOCs) such as unburned hydrocarbons, carbon monoxide, NO and NO_2 (collectively called NO_x), and SO_x. Generally, important REDOX reactions in a catalytic converter can be broken into three categories related to the exhaust gases they address: NO_x reduction, volatile organic carbon (VOC) and carbon monoxide (CO) oxidation, and other types of REDOX reactions.

NO_x contributes to the typical urban smog cycle and gives smoggy air the brownish coloring one sees over cities where the climate and geography are conducive to smog formation. Your catalytic converter helps minimize this brownish haze by reducing the NO_x in the exhaust stream to nitrogen gas and oxygen-bearing species in reactions that use hydrogen, carbon monoxide, or unburned hydrocarbons as the reducing agent. The three most common electron-transfer reactions in the catalytic converter that accomplish NO reduction are:

$$2NO\ (g) + H_2\ (g) \rightarrow N_2\ (g) + H_2O\ (g)$$

$$2NO\ (g) + 2CO\ (g) \rightarrow N_2\ (g) + 2CO_2\ (g)$$

$$38NO\ (g) + 2C_6H_{14}\ (g) \rightarrow 19N_2\ (g) + 12CO_2\ (g) + 14H_2O\ (g)$$

In these reactions, each nitrogen atom in the NO acquires two electrons to produce N_2. The last reaction is just one representative case of a NO/unburned-hydrocarbon REDOX reaction; any hydrocarbon could replace the hexane on the reactant side of this reaction and will simply result in an alteration of the reaction stoichiometry. The chemical reactions leading to NO_2 reduction are quite similar; however, the reaction stoichiometry also changes, since four electrons must be transferred to each nitrogen to fully reduce it to N_2 (g). This series of possible reactions for NO_x reduction has already introduced two catalytic converter reactions related to the second major objective of catalytic converters, namely the oxidation of CO (g) and unburned hydrocarbons. However, catalytic converters also promote the oxidation of alkenes and aldehydes to carbon dioxide and water in the presence of O_2 (g). Recall that alkenes are organic compounds with at least one C=C double bond and that aldehydes all contain carbonyl groups (C=O), where there are either one or two hydrogen atoms directly bonded to the carbonyl carbon atom (see Appendix C). Other typical oxidation reactions in the catalytic converter include:

$$2CO\ (g) + O_2\ (g) \rightarrow 2CO_2\ (g)$$

$$C_nH_m\ (g) + (n + m/4)O_2\ (g) \rightarrow nCO_2\ (g) + (m/2)H_2O\ (g)$$

$$CH_2O\ (g) + O_2\ (g) \rightarrow CO_2\ (g) + H_2O\ (g)$$

As in the earlier example, any unburned hydrocarbon can be used on the reactant side of the second reaction, and any aldehyde can substitute for formaldehyde (CH_2O) in the final reaction. Catalytic converters also perform other functions in addition to these two primary REDOX reactions. One important reaction in this "other" category is the reduction of sulfur dioxide gas by water vapor in the exhaust stream, which reduces sulfur from the +4 to the –2 oxidation state via a six-electron transfer process:

$$2SO_2 \text{ (g)} + 2H_2O \text{ (g)} \rightarrow 2H_2S \text{ (g)} + 3O_2 \text{ (g)}$$

H_2S (g), while not completely inert, plays a lesser role than SO_2 (g) in the standard atmospheric processes that lead to acid rain.

The catalyst for these reactions is often an inert metal, such as platinum, which can easily adsorb gases in the exhaust stream and facilitate interactions between gas molecules to produce new chemical species. Adsorption (or physisorption) is the process by which a molecule forms a close association with a surface, but the association is based on intermolecular forces between discrete molecular units rather than chemical bonding (Figure 3.3). Physisorbed materials are weakly held by a surface and can easily migrate laterally along that surface via the process of diffusion. Importantly, it takes relatively low energy to desorb a physisorbed molecule from another entity, since you do not need to break any chemical bonds. The platinum and other rare earth metals in catalytic converters are types of heterogeneous catalysts, or catalysts that are not in the same state of matter as the reactant molecules (solid versus gas in this case). But if the catalyst is a solid and does not participate in the reaction itself, how does it help speed up gas-phase reactions in your exhaust stream? One way is by exploiting the properties of physisorption: The reactant gases adsorb to the catalyst surface and can either diffuse toward one another along the surface or sit in one place and encounter a greater number of other reactant gases as they speed past. One reactant gas molecule being sorbed to the stationary surface increases the mean relative speed for collisions with other gas molecules because the adsorbed species is

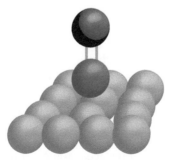

Figure 3.3 (See color insert.) Physisorption of carbon monoxide at the surface of a platinum metal. The dark sphere represents carbon, the red sphere oxygen, and the light gray spheres the atoms of platinum at the platinum cluster surface. Note the lack of chemical bonds between the carbon and the surface.

no longer translating rapidly through the exhaust system. Basically, both scenarios increase the probability of an interaction occurring between the desired gas molecules with the right energy and/or molecular orientation. However, this basic discussion presents an overly simplified view of the catalytic process, as one study has shown that some reaction mechanisms in catalytic converter chemistry involve up to 61 elementary steps.

One potential problem with modern catalytic converters is that they need to operate at rather high temperatures to function efficiently. The typical operating temperature of a modern catalytic converter is in the range of 300°C and above, where conversion efficiency is 90% and greater. It is nearly impossible to achieve 100% conversion efficiency in a catalytic converter, primarily because the dominant types of catalytic reactions change as the fuel/air ratio and amount of unburned fuel vary. Likewise, competition among exhaust gases for the limited number of catalyst sorption sites also reduces practical efficiency. Beyond these factors, the catalyst performs its many functions rather poorly at low temperatures. This can be rationalized to some degree using the concept of activation energy, but in this case, the activation energy most likely has to do with establishing an appropriate rate of surface diffusion or a high mean relative speed. Most (≈80%) of the harmful emissions generated by a typical internal combustion vehicle occur during the first minute of operation, when the exhaust temperature is too cold to promote efficient catalyst performance. In an effort to improve cold-start efficiency, some manufacturers have begun to explore the possibility of preheating catalytic converters, though this technology does not currently find wide application. Others have taken the step of relocating the catalytic converter closer or adjacent to the exhaust manifold, where it will contact hotter exhaust gases and reach a useful operating temperature more quickly.

Since direct contact between the catalyst particles and the exhaust stream is required for the catalytic converter to function, the material design of a catalytic converter attempts to protect the platinum catalyst as much as possible, facilitate a high contact area between catalyst particles and the exhaust gases, and maintain structural stability under the harsh and variable conditions characteristic of the hot exhaust gases. At the heart of a catalytic converter is a ceramic support structure that holds the inert metal catalyst (Figure 3.4). Often, this ceramic support has a honeycomblike design, which provides minimum restriction for the exhaust gas flow while maintaining a high-contact surface area. The platinum or other inert metal catalyst is strongly attached to this ceramic support near the solid–gas interface. This ceramic assembly is wrapped in an expandable sheath and placed within an insulating metal container. Ceramics are an ideal material for the catalyst support, as they often have low thermal expansion coefficients, preventing rupture of the converter body at the high operating temperatures.

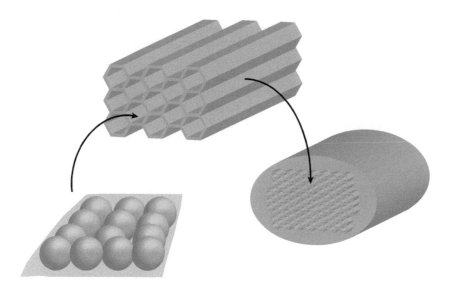

Figure 3.4 (See color insert.) Platinum and other transition metal catalyst particles are embedded at the surfaces of a honeycomb-like ceramic support in the catalytic converter.

There are several mechanisms by which a catalytic converter can become damaged or ineffective at promoting its important REDOX reactions. The biggest chemical concern is the concept of fouling, or loss of catalytic activity due to an irreversible chemical process. In a catalytic converter, chemical fouling occurs when some compound or element irreversibly adsorbs to the catalyst surface, permanently occupying a site that could otherwise be used for a catalytic reaction. When lead and methyl tert-butyl ether (MTBE) were used in fuel, both of these materials could irreversibly sorb or coat the catalyst surface, preventing the exhaust gases from reaching sorption sites. In modern fuels, the most prominent chemical fouling agent is SO_2 gas, which sorbs very strongly to the metal surface and is difficult to remove. It is this same metal–sulfur chemistry that drives the use of alkane thiol molecules to functionalize gold nanoparticles. This particular type of chemical fouling is most troublesome for vehicles that burn high-sulfur fuels, and many fuels of this type are no longer used because of their role in acid rain formation.

3.4   Rust and corrosion

Chemistry Concepts: REDOX reactions, electrochemistry
Expected Learning Outcomes:
- Explain the chemical reactions that lead to rust formation
- Understand why salted roads promote the formation of rust

The rusting of iron-bearing parts and other forms of corrosion are also electrochemical processes, and the loss or failure of components due to corrosion reactions represents a significant cost to our economy each year. Corrosion is generally defined as the gradual destruction of a material due to chemical reactions with the material's local environment, and this encompasses many more reactions than simply the rusting of steel or oxidation of other metals. Corrosion has been widely studied in many different disciplines, and makes up the subject matter for a large number of books, courses, and ongoing research projects. Clearly, it is impossible to fully discuss even the corrosion of metals here.

However, in the case of automotive corrosion, rust formation is perhaps the most important corrosion process and involves electrochemical reactions between the iron in cast iron and steel parts with oxygen and water. The anodes and cathodes in this reaction are the iron and the oxygen gas, respectively. In the presence of excess water and limited oxygen, as we would expect beneath a water droplet on the surface of your car, the associated reactions are:

$$Fe \text{ (s)} \rightarrow Fe^{2+} + 2e^- \qquad E^0 = -0.447 \text{ V}$$

$$4H^+ + O_2 \text{ (g)} + 4e^- \rightarrow 2H_2O \qquad E^0 = 1.23 \text{ V}$$

The presence of extra oxygen and an acidic water droplet can also drive the Fe(II) to Fe(III) as follows:

$$4Fe^{2+} + 4H^+ + O_2 \text{ (g)} \rightarrow 4Fe^{3+} + 2H_2O \qquad E^0 = -0.771 \text{ V}$$

These REDOX reactions release ions into the water droplet that convert the droplet into the electrolyte solution that completes the electrochemical cell. Once the iron is ionized via the REDOX chemistry described here, it can undergo a number of possible non-REDOX reactions with the water to form iron hydroxide precipitates and generate the acid protons required for the oxygen reduction reaction:

$$Fe^{2+} + 2H_2O \text{ (l)} \rightarrow Fe(OH)_2 \text{ (s)} + 2H^+$$

$$Fe^{3+} + 2H_2O \text{ (l)} \rightarrow FeO(OH) \text{ (s)} + 3H^+$$

$$Fe^{3+} + 3H_2O \text{ (l)} \rightarrow Fe(OH)_3 \text{ (s)} + 3H^+$$

These iron hydroxides can undergo subsequent dehydration reactions (reactions where an H_2O molecule is eliminated from the structure) or additional reactions with dissolved oxygen gas to produce a series of solid oxyhydroxides and oxides that we commonly recognize as rust. One possible path from Fe^{3+} to iron(III) oxide (red rust) is as follows:

$$Fe^{3+} + 3H_2O \text{ (l)} \rightarrow Fe(OH)_3 \text{ (s)} + 3H^+$$

$$Fe(OH)_3 \text{ (s)} \rightarrow FeO(OH) \text{ (s)} + H_2O \text{ (l)}$$

$$2FeO(OH) \text{ (s)} \rightarrow Fe_2O_3 \text{ (s)} + H_2O \text{ (l)}$$

Alternatively, iron(II) hydroxide can directly convert to a hydrated form of iron(III) oxide in the following REDOX reaction:

$$4Fe(OH)_2 \text{ (s)} + O_2 \text{ (g)} \rightarrow 2Fe_2O_3 \cdot H_2O \text{ (s)} + 2H_2O \text{ (l)}$$

Other mechanisms exist for the conversion of oxidized iron to red rust, but interested readers are left to investigate these mechanisms further on their own.

Clearly, the role of water in automotive corrosion is critical and influences the formation of rust in several important ways. First, you can see that water is a key reactant in all the non-REDOX reactions involving aqueous iron ions, and it is clear that water is also a product in many of the other reactions involved in red-rust formation. Second, the excess water in a water droplet serves as a solvent and ultimately becomes the electrolyte solution that completes the electrochemical cell between the iron and oxygen gas. Clearly, the conductivity of the droplet, or the ease with which an electrolyte can transport a charge, will play a role in the speed at which a material rusts. The conductivity of the water droplet is influenced by the ions generated in the REDOX reactions and other external factors. For example, acid rain droplets accelerate rusting because they are naturally strong electrolyte solutions due to the acid protons and the nitrate and sulfite/sulfate anions they carry. Likewise, road salt will also promote corrosion of iron and steel by providing an additional source of ions in the water droplet. Indeed, corrosion of automobiles is often most severe in locations where salting of roadways is required to control ice formation in the winter. Finally, the water plays a key role in rusting by trapping oxygen and slowing the molecular motion of dissolved oxygen, leading to a longer surface residence time that facilitates these corrosion reactions. Even in the presence of efficient electron transfer, the rusting of iron is significantly slower when the iron must react with the fast-moving molecules of O_2 (g) and H_2O (g) in the atmosphere. Thus, the goal of corrosion protection in the automotive industry is to prevent water from coming into contact with steel and cast iron parts, especially those that affect the aesthetics or critical mechanical functions of the car.

Corrosion processes almost always involve at least one REDOX reaction, and perhaps not surprisingly, REDOX chemistry and electrochemistry are both used to protect vehicles against corrosion. In the remainder of this section, we provide a brief overview of the most common methods of preventing corrosion. The interested reader is referred to the review article by Hannour and Rolland (2009) and the references therein for more detailed discussions of anticorrosion chemistry.

Typically, steel body panels are provided several different layers of corrosion protection that rely on various types of chemical reactions, not all of which involve electron transfers. For example, some of the most obvious examples of nonelectrochemical corrosion protection are painting, clear coating, and waxing. Each of these methods is designed to provide a coating impenetrable to water on the surface of the body panels or rust-prone parts of the vehicle. However, paints, clear coats, and filler coats are very thin coatings, and their protective layer can easily be penetrated during accidents or even by small impacts from rocks and debris on the road. Waxing helps to keep the surface protected even when the paint has been damaged by providing a thin, hydrophobic (water-repelling) coating that prevents wetting of the surface and water from settling into scratches and dings. However, waxes are so easily removed that it is recommended you apply them at least twice a year.

Below the thin coats of paint and filler is the first type of electrochemical protection, called the *electrocoat*. This is a roughly 10–30-micron polymer coating that is applied by passing the panel through an aqueous suspension of charged polymer while applying an external potential to the panel. Some type of electrode is also present in the bath to complete the electric circuit. The main electrochemical reaction in aqueous electrocoating is water electrolysis, and the electrocoating process is considered to be either anodic or cathodic, depending on the electrolysis half-reaction occurring at the body panel:

$$\text{Anode: } 2H_2O \rightarrow 4H^+ + O_2 \text{ (g)} + 4e^- \qquad E^0 = -1.23 \text{ V}$$

$$\text{Cathode: } 2H^+ + 2e^- \rightarrow H_2 \text{ (g)} \qquad E^0 = 0.00 \text{ V}$$

Although any polymer that participates in this process must be able to carry a stable charge, the types of charged polymer that can be used in this process are quite diverse, and polymers are available for both cathodic and anodic deposition processes. The polymers themselves offer no electrochemical protection against corrosion of the metal panel, but they do provide a second and more durable hydrophobic coating designed to prevent oxygen and water from making contact with steel or cast iron. A significant drawback to the electrocoating process is that it adheres rather poorly to metal surfaces and can reveal the grain structure of the metal (for an example of graining, look at the metal ductwork in your home heating system), and it thus requires a preparative step prior to the electrocoating.

The most common preparation for electrocoating is to phosphatize the surface which converts the surface layer of the material to a metal phosphate phase. Phosphatization essentially requires exposing the metal surface to a solution of phosphoric acid, phosphoric acid salts, other metal phosphates, or a combination of these with other additives that then lay a strongly adhering metal phosphate coating on the panel. A generic reaction for the phosphatization of steel is:

$$Fe\ (s) + M^{3+} + 2H_3PO_4 \rightarrow FePO_4 + MPO_4 + 3H_2\ (g)$$

where M^{3+} is any trivalent metal ion. Here, the iron at the panel surface is oxidized to Fe^{3+}, which can then form the phosphate phase while acid protons are reduced to hydrogen gas. In the automotive industry, one or more metal ions other than iron(III) are often available in the solution, producing a very thin mixed-metal phosphate coating. Prior to electro-coating, metal panels will be dipped in the phosphatization solution and held there until production of hydrogen gas stops. Once again, the phosphatization layer provides little electrochemical resistance to corro-sion and, according to Hannour and Rolland, is mainly used to facilitate strong and stable adhesion of the electrocoating.

The most important protection against iron corrosion is the galvani-zation process, which involves coating an iron or steel part in zinc. Recall that galvanic cells are electrochemical cells, where the associated REDOX reactions occur spontaneously. In galvanization, the metal part you wish to protect from corrosion must be coated with a metal that is more eas-ily oxidized. Then, when the panel is exposed to conditions conducive to corrosion, the more easily oxidized metal (in steel galvanization, the zinc) functions as a sacrificial metal that will be oxidized first via a more spontaneous process, thereby providing a true electrochemical method of protection. This is precisely the function of the zinc in steel galvanization, as shown by the higher positive voltage for oxidation of zinc versus iron:

$$Zn\ (s) + 2e^- \rightarrow Zn^{2+} \qquad E^0 = +0.76\ V$$
$$Fe\ (s) + 2e^- \rightarrow Fe^{2+} \qquad E^0 = +0.41\ V$$

When the electrolyte solution comes into contact with the zinc, acidic pro-tons dissolve the zinc metal via a REDOX process to produce precipitates of zinc oxide, zinc hydroxide, and zinc chlorides that coat the steel and provide additional protection for the underlying steel:

$$Zn\ (s) + 2H^+ \rightarrow Zn^{2+} + H_2\ (g)$$
$$Zn^{2+} + 4ZnO + 4H_2O + 2Cl^- \rightarrow ZnCl_2 \cdot 4Zn(OH)_2$$
$$Zn^{2+} + 8H_2O + 2Cl^- \rightarrow ZnCl_2 \cdot 4Zn(OH)_2 + 8H^+$$

These reactions tend to produce a white to grayish-white color. When the zinc layer is fully consumed, or when a scratch or ding penetrates all of the protective layers down to the steel itself, then iron corrosion reactions begin. The most common modern method of galvanizing steel panels is the hot-dip galvanization process. After several cleaning and degreasing steps, a steel panel is covered with a flux that protects the cleaned steel from oxidizing in air and facilitates better wetting with the zinc metal. The flux-coated panel is then dipped into a bath of molten zinc until the temperature of the steel part has equilibrated with the temperature of the bath. The panel is then removed and rapidly quenched to complete the galvanization. Generally, this process generates a zinc layer roughly 10-μm thick on the surface of the panel.

Another approach to eliminating rust on vehicles is to use metals other than iron or steel, or to replace parts with nonmetallic structural materials like plastic or carbon fiber. However, the use of other metals such as aluminum is only a partial solution, since aluminum is subject to corrosion reactions of its own. Contact points between different metals that are wet also form electrochemical cells, which can promote galvanic corrosion of the more easily oxidized metal. We discuss the use of advanced hybrid materials for vehicles in Chapter 6 and leave readers to pursue corrosion of aluminum on their own.

Anticorrosion Additives in Fuels

Just like the exterior of a car exposed to salty road solutions in the winter, the hot environment of the combustion chamber is another place susceptible to corrosion. Clearly, corrosion-related damage in the engine is very undesirable, and chemists have devised several chemical additives for fuels that help to limit corrosion during and after combustion. One such example is zinc dithiophosphate chemicals, which provide many of the same anticorrosion properties inside the combustion chamber that we apply to external body panels. Commercially, these additives are often known by the acronym ZDDP, which stands for zinc dialkyldithiophosphates. These chemicals are not salts, but rather coordination compounds that involve direct bonding of the zinc with the sulfide anions of the dithiophosphate functional groups in a tetrahedral coordination. In other words, the zinc is linked to four sulfide sulfur atoms such that all Zn–S bonds are ≈109.5° from the other three Zn–S bonds in three-dimensional space. The variably sized alkyl chain allows one to make the ZDDP soluble in either oils or gasoline/diesel fuel by tailoring the alkylphosphate chemistry (see Chapter 4, Intermolecular Forces). Typically, combustion of ZDDP contained in the oil or fuel generates a thin coating of sulfides on the surfaces of valves and other metal components in the engine in addition to providing some degree of phosphatization and zinc coating on the metal surfaces. These coatings are then subjected to and/or sacrificed during the REDOX reactions in the combustion chamber, protecting the metallic surfaces of the valves, block, and heads from corrosion-related engine wear.

3.5   Chrome plating

Chemistry Concepts: electrochemistry, REDOX reactions, electrolytic reactions, activity series

Expected Learning Outcomes:
- Explain the chemical process of chrome plating using the REDOX activity series
- Define and describe the process of electroplating in general

As mentioned in the previous section, plating with a metal more susceptible to chemical attack than the one you wish to protect is one way to prevent corrosion. It has also been traditional in automobiles to use plating of both metals and plastics with nickel, chrome (chromium), and sometimes other metals for cosmetic reasons. Plating is an electrochemical process that often involves many chemicals, REDOX reactions, applied electric currents, thorough rinsing and cleaning, and both simple and advanced or proprietary catalysts to accomplish the overall goal of laying a thin metal coating on a part. While nickel plating is also employed in automotive applications, in this section we will focus on the historical and modern approaches to chrome plating in automobiles.

Chromium plating is accomplished via an electrodeposition process, which is essentially an electrolytic process similar in nature to the electrocoating process described in the previous section. The substrate (the material that will be plated) is electrified by passing a current through it. In the case of chromium plating, the substrate must act as the cathode, providing additional electrons so that the chromium in solution can be reduced to chromium metal on the surface. An additional electrode of some type must also be present in the bath to complete the circuit.

There are two types of chromium plating: decorative and functional. Decorative plating typically has a deposited layer thickness of <0.80 micrometers, while "hard chrome" or functional chrome exceeds this layer thickness and imparts both aesthetic and mechanical properties to the substrate. In both cases, the chromium layer is deposited either directly on the substrate, or more often, the substrate is coated with a nickel plate before chromium is applied. Chrome plating, particularly functional chrome plating, takes advantage of the properties of chromium, including its resistance to corrosion, its ability to form a stable and self-healing oxide layer, hardness, heat resistance, and durability or mechanical strength. Catalysts are required during the chromium plating process, primarily to increase the conductivity of the plating solution so that a thin, uniform film can be produced easily.

Traditional chrome plating involves the use of Cr^{6+} in solution and is generally called the hexavalent chromium process. A hexavalent chromium bath is typically prepared by dissolving chromic anhydride, CrO_3, in water. This material is sometimes called chromic acid because the anhydride forms chromic acid (H_2CrO_4) in water, which readily polymerizes via a dehydration reaction to form $H_2Cr_2O_7$ (known as dichromate) and larger polyacids. Most commercial baths typically contain 200–400 g/L of chromic anhydride. A source of additional anions is also required to facilitate effective chromium plating, and often this is provided as sulfate from sulfuric acid or sodium sulfate or fluoride from fluorosilicic acids or silicofluoride chemicals. It is currently believed that these anions help form acid radical species that catalyze the Cr^{6+} deposition process, and indeed the concentration of anion has been shown to exert a strong influence on the deposition rate. The first step in a chrome plating process is to clean the metal surface thoroughly, because any dirt, oil, or other contamination will make it difficult for the chromium to reach the material surface and form a strong coating. Occasionally, the substrate will also be placed in a chromic acid activation bath, where the current in the substrate causes it to act as an anode, which results in the chromic acid etching the substrate surface and removing any last contaminants from the metal. Preplating with other metals like copper and nickel may then be performed. Finally, the substrate will be charged to act as the cathode and put into the chromic acid plating bath. Here, the overall process involves the substrate surface providing the electrons to reduce chromium(VI) in dichromate to chromium metal in an acidic solution:

$$Cr_2O_7^{2-} + 14H^+ + 9e^- \rightarrow 2Cr\ (s) + 7H_2O\ (l) \qquad E_0 = +0.59\ V$$

In practice, voltages of 4–12 V are required for effective chromium electroplating. After being exposed to the plating solution, the newly plated substrate must go through several cleaning baths that will remove any remaining plating solution. The solution temperature and the current density in the Cr(VI) plating bath are critical to forming the desired product. Temperatures between 35°C and 55°C and current densities of

3–26 A/dm^2 will produce "bright plate" required to produce the polished and highly reflective surfaces. Details of the plating chemistry are quite complicated, and the interested reader is referred to Chapter 7 of *Modern Electroplating* for a detailed description.[*]

There are several drawbacks to the hexavalent chromium process, not the least of which is the toxic nature of hexavalent chromium. Cr^{6+} is a known carcinogen (cancer-causing chemical) and is an acute poison in the environment. In addition, the pH of most hexavalent chromium plating baths is on the order of pH = 0 or less, making the bath extremely corrosive. Both the toxicity and corrosive nature of the solution make hexavalent chromium bath wastes hazardous, increasing the disposal costs and making plating less economical.

Newer chromium plating processes make use of chromium(III) plating chemistry, which reduces the environmental hazards. Chromium(III) is spontaneously generated from chromium(VI) in the presence of excess electrons in solution, meaning no special conditions are required to produce chromium(III):

$$Cr_2O_7^{2-} + 14H^+ + 6e^- \rightarrow 2Cr^{3+} + 7H_2O(l) \qquad E_0 = +1.33\,V$$

The trivalent state of chromium is capable of producing a wide array of complex cations that also produce a wide array of colored solutions. For example, $[Cr(H_2O)_6]^{3+}$ readily forms in solution and generates a violet color due to splitting of the chromium d-orbital energies. For the trivalent chromium processes, the sulfate and fluoride anions are used to form chromium complexes that facilitate chromium electrodeposition. The wide variety of complex cations and polymeric cations that chromium(III) can form make describing the detailed chemistry beyond the scope of this text. Chromium(III) is not easily reduced to chromium metal, and a current through the substrate is required to form the chromium metal coating:

$$Cr^{3+} + 3e^- \rightarrow Cr\ (s) \qquad E_0 = -0.74\ V$$

The actual plating baths and the electroplating process are quite similar to the hexavalent chromium approach in other respects, and thus there is no need to repeat that discussion here.

---

[*] N. V. Mandich and D. L. Synder, "Electrodeposition of Chromium," in *Modern Electroplating*, 5th ed., ed. M. Schlesinger and M. Paunovic. (New York: John Wiley and Sons, 2010), chap. 7.

chapter four

Intermolecular forces

Often, macroscopic observations made with one's senses are rooted in the types and strengths of attractions/repulsions between chemical species on a molecular scale. For example, your eyes tell you that waxing a car causes water to form large droplets called beads, but how does it do this, and how does wax protect a car body from rusting? Why do soaps foam and why are there soaps and specialty automotive detergents for your car? Why are there different "weights" for motor oil and why does your car require a particular one? All of these questions are related to intermolecular (IM) forces, or the forces that develop between molecules in vapors, liquids, and at solid surfaces. It is important to distinguish in your mind the difference between bonding interactions, which occur within a molecule, and the forces between molecules. For an individual atom and for chemical reactions, the bonding forces within molecules are important. It is the energy stored in chemical bonds that make hydrocarbons like octane excellent fuels for your engine. For understanding large collections of molecules that we can see, hold, and touch, intermolecular forces become more important. For example, the very strong intermolecular forces between H_2O molecules in water are the reason behind the anomalously high boiling point for this small molecule. In this chapter, we will introduce the types of intermolecular forces and how they manifest in the chemistry of your car.

4.1 Types of intermolecular forces

Chemistry Concepts: intermolecular forces, organic chemistry
Expected Learning Outcomes:
- Name the common types of intermolecular forces
- Understand the origin of the various forces
- List some common material properties that depend upon IM forces

Molecules are typically considered to be atoms held together in fixed atomic arrangements via chemical bonds, but they can also be considered collections of protons, neutrons, and electrons with relatively well-defined spatial relationships. Sometimes these arrangements of protons, neutrons, and electrons are not symmetrically distributed in space, leading to molecular dipole moments that cause individual molecules to function on some basic level as small magnets. These asymmetric distributions of particles/charge also give rise to several types of electronic interactions, such as electrostatic attractions between regions of partial positive and partial negative charge in an H_2O molecule. Most chemistry students have a fairly well-developed idea about molecular structure, molecular dipole moments, and properties of molecules in isolation by the end of general chemistry, but what about understanding what happens when two molecules approach one another in space? For example, how and why does a droplet of water stay together on the surface of your car rather than dispersing completely? The answer is that molecules can and do interact via electronic and magnetic attractions/repulsions that are too weak to form chemical bonds. It is these relatively weak attractive and repulsive forces that make up the intermolecular forces we discuss in this chapter.

There are several mechanisms by which molecules can interact, and each mechanism has its own special set of circumstances that must exist for the interaction to be significant. One type of intermolecular force is the *dipole–dipole interaction*. This occurs exclusively when two molecules that have net dipole moments approach one another in space. In other words, this interaction is important only for collections of polar molecules, which are molecules that have an asymmetric distribution of electron-rich and electron-poor regions. The resultant spatial variation (or gradient) in negative charge generates both an electric and a magnetic dipole moment that points from the electron-poor region toward the electron-rich region of the molecule. Both the electric and magnetic dipole moments contribute toward the overall dipole–dipole interaction. The electric dipole moments interact via an electrostatic attraction between the electron-depleted region of one polar molecule and the electron-rich region of another—an instance of opposite charges attracting, much like we observe during the formation of an ionic bond. However, in this case, the distances between the charged regions are larger and the charges much smaller, making the electric dipole interaction a relatively weak intermolecular force rather than a strong bonding-type interaction. The electrostatic-attraction component typically dominates the dipole–dipole interaction in the case of small polar molecules, but there is also a generally less important magnetic component to the interaction that arises when the net magnetic dipole moments are arranged such that an attractive force develops. Though ferromagnetism is different from the magnetic and electric interactions discussed here, insight into factors affecting the overall strength of the dipole–dipole interaction can be gained by considering what happens when you place the negative and positive poles of two ferromagnets close together. You have to work to keep the two magnets apart, and the effort required to keep them apart must exceed the attractive force between the magnets. The smaller the distance between the two poles, the more effort is required to keep the poles apart. In other words, a shorter distance between the poles equals a stronger attractive force. The bigger the magnets, the greater the force required to keep the poles apart at a fixed distance. Likewise, dipole–dipole forces between molecules are stronger when the dipole moments of the molecules are larger, equivalent to having a more powerful bar magnet. Strong dipole–dipole forces also develop when the molecules themselves are more compact in a three-dimensional sense, which lets the molecules get closer together. Though the bar-magnet analogy does not involve charges, the distance between the partial charges and their magnitudes affect the electrostatic attraction component of the dipole–dipole interaction in the exact same fashion. Thus, simply by arranging the molecules such that the positive end of one molecular dipole is near the negative end of another, we develop an attractive dipole–dipole interaction in any system of polar molecules.

A second important intermolecular force in systems that contain polar molecules is the *ion–dipole interaction*, in which an ion is attracted to one of the poles on the polar molecule. Recall that dipole moments develop from nonuniform distributions of charge in space. As discussed in the previous paragraph, any molecule with a dipole moment contains a region that has a net partial negative charge and a net partial positive charge. If you have a partially negative region in a molecule and it approaches a positively charged ion, then an electrostatic attraction develops, because unlike charges attract. An analogous line of reasoning can be used to explain the attraction of a negative ion to the positive region of a polar molecule. Ion–dipole forces can only exist if a system contains both ions and polar molecules, such as the situation in saltwater. Ion–dipole forces are similar in magnitude to dipole–dipole forces and are stronger when the ion has a greater charge, when the ion has a smaller size, when the molecule has a larger dipole moment, or when the molecule has a smaller size.

Another very important intermolecular force that is present only in specific situations is *hydrogen bonding*. Hydrogen bonds are too weak to be considered chemical bonds, but they are much stronger than the other types of intermolecular forces, typically an order of magnitude stronger (e.g., 40 kJ/mol versus 4 kJ/mol for dipole–dipole interactions). Currently, it is accepted that hydrogen bonds form when hydrogen bound to oxygen, fluorine, or nitrogen interacts with lone-pair electrons on another oxygen, fluorine, or nitrogen. When the hydrogen bond donor (the hydrogen atom in an N–H, O–H, or F–H bond) and acceptor (an N, O, or F with a lone pair) are on different molecules, then hydrogen bonding becomes a very strong intermolecular force. Hydrogen bonds hold together your DNA and are responsible for many of the unusual properties of water, such as the fact that liquid water is denser than solid ice.

But what about when the molecules do not have dipole moments, cannot form hydrogen bonds, and do not form ions? For example, the *n*-octane in gasoline can do none of these things, yet gasoline is a liquid, which implies that the molecules of *n*-octane are interacting with each other to a relatively significant extent; otherwise, *n*-octane would be a vapor at normal atmospheric conditions. The explanation lies in the fourth major type of intermolecular force that is present in every chemical system, called *dispersion forces*. Dispersion forces include the traditional dipole-induced dipole, ion-induced dipole, and induced dipole-induced dipole interactions. Essentially, dispersion forces are the very short-lived attractions that form when two molecules approach one another and their electrons interact (Figure 4.1). We all know that electrons move and that they move very quickly. Typically, they move so quickly that we do not concern ourselves with exactly where they are located at any given moment. But if we could freeze an atom or molecule in time, we may find that more electrons are found on one side than another. This effectively creates what is called an *instantaneous dipole*, meaning that it exists for a very, very brief moment in time. Now, if there is a second molecule near the temporarily electron-depleted side of our first molecule, the electrons of the second molecule feel a stronger pull from the positive charge of molecule one's nucleus. The result is a very brief and fairly weak attractive interaction between molecule one and molecule two. Though these attractions are very short lived, in a chemical system with 10^{23} atoms, an enormous number of these weak attractions form every second, leading to a fairly strong total attractive force between otherwise neutral and nonpolar molecules. Dispersion forces become stronger when molecules have a less compact shape and when molecules are larger—essentially opposite the trends for all of the other intermolecular forces we have discussed thus far.

If you are having trouble imagining dispersion forces, picture what happens on a crowded bench seat at a sporting event. Let's say you are on the end of the row and that at the current row occupancy there is just enough space for everyone to face directly forward so that everyone's shoulders touch. Then one more person sits down in the row. Clearly, there is not enough space for that person's shoulders, so they turn slightly to one side. This leaves you with two choices: you can continue to face directly forward and put up with the squeezing caused by the lack of space, or you can turn you shoulders in the same direction to reduce the squeezing force. Atoms and molecules do the same thing with their electron distributions: When the number of electrons in a space gets too crowded, the molecules adjust their distribution of electrons to make everything more comfortable, leading to lower energy in the system and our attractive dispersion force.

Intermolecular forces are responsible for several important properties of fluids used in a car. Viscosity, surface tension, boiling points, and melting points are all direct results of the types and strengths of intermolecular forces present in a chemical system. They also influence the solubility of materials. We discuss these topics in the remainder of this chapter.

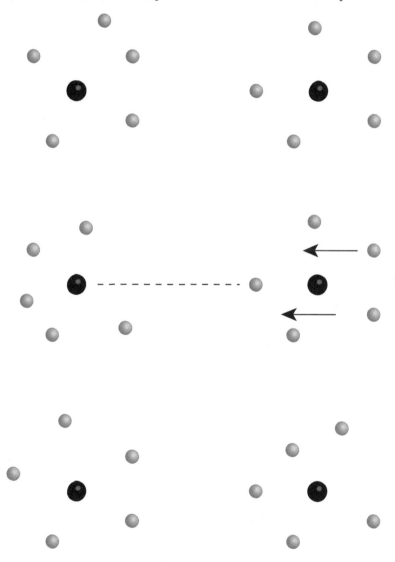

Figure 4.1 Dispersion interactions. The black spheres represent the nuclei, and the gray spheres denote electrons circulating about each nucleus. In the second image, the nonsymmetric arrangement of electrons on atom #1 causes a brief attraction between the nucleus of atom #1 and the electrons of atom #2.

4.2   Solubility

Chemistry Concepts: solution chemistry, solubility, intermolecular forces
Expected Learning Outcomes:
- Explain the concept of solubility and solubility limits
- Use intermolecular forces to explain the relative solubility of various materials

In the previous section, we briefly mentioned saltwater as an example of a system involving ion–dipole forces. Saltwater is a simple mixture of salt and water. If you take table salt and put it in water, you will see the salt crystals vanish as saltwater forms. Furthermore, if you drink some of the saltwater from the top of the glass, then drink some from the bottom of the glass using a straw, they will taste exactly the same if the drink was well mixed. This tells you that there must be both salt and water in every part of the saltwater. The saltwater is an example of something called a solution, and intermolecular forces play important roles in solutions and the solubility of materials, or the ease with which two substances can form a solution. Many parts of a car (liquid, solid, and vapor) are solutions, and thus a discussion of solutions and how they form is essential to understanding car chemistry.

Technically speaking, a solution is a mixture of two or more chemical components such that all components are evenly distributed throughout the volume and one component dominates the composition. A mixture with evenly dispersed chemical components is often called a homogeneous mixture. The component present in the largest amount is called the *solvent*, and any components present in lesser amounts are called *solutes*. In our salt-water example, the water is the solvent and the salt is the solute. Solutions can occur in solids, gases, or liquids. For example, steel is a solid solution of iron and carbon, where iron is the solvent and carbon is the solute. Air is an example of a gas-phase solution, with oxygen as a solute in a solvent of nitrogen. When two or more components can form a homogeneous solution, we say that the solute is soluble in the solvent, or that the solute can dissolve in the solvent. Dissolution, or dissolving, is the process in which the atoms/molecules of the solute are completely surrounded by solvent molecules and removed from the pure solute phase. The term dissolution is typically understood to describe the process in which a solid solute dissolves in the liquid solvent, like our salt water example. The dissolution process is highly dependent on the properties of the solute and solvent and the types of intermolecular forces that can exist between solute and solvent.

So why do the salt crystals dissolve when we make saltwater, but persist if we pour salt into vegetable oil? Table salt, sodium chloride (NaCl), is an ionic compound in which the cations and anions are held together via very strong electrostatic attractions between positively and negatively charged ions. Water is a small polar molecule with a relatively large partial negative charge on the oxygen atom and a relatively large partial positive charge on the end with the hydrogen atoms. When water approaches a sodium ion at the surface of the sodium chloride salt crystals, the oxygen in the water orients toward the sodium, forming a strong ion–dipole interaction and partially neutralizing the charge of the sodium ion. When several water molecules interact with the sodium ion in this way, the sodium ion is held more weakly to the solid salt crystal. In other words, its electrostatic attraction to the chloride anions in the crystal is becoming nearly equal to its electrostatic attraction to the negative end of the water dipoles in the solvent. Eventually, the sodium ion will leave the salt surface and become completely surrounded by water molecules, with the resulting ion–dipole interactions leading to charge neutrality. To put it more simply, the fact that water can participate in ion–dipole interactions and that sodium and chloride ions can participate in ion–dipole interactions means that the salt can dissolve in the water. When the salt crystals are placed in vegetable oil, they do not dissolve. This is because the vegetable oil molecules are predominantly nonpolar organic molecules, meaning that nothing other than dispersion interactions are present between the salt crystals and the oil solvent. These interactions are not strong enough to overcome the electrostatic attractions between the ions in the salt crystal, making the salt insoluble in the oil. This example illustrates an important rule about solubility and intermolecular forces that is typically summarized as "like dissolves like." In other words, solutions form only when similar types and strengths of intermolecular forces exist between the solute and the solvent.

Another important concept in solubility is the idea of solubility limits. Sometimes, large amounts of solute can dissolve in a solvent, while in other instances, solutes are only somewhat soluble in solvents. To better understand this idea, picture a group of 10 people picking up boxes. Each person can carry 50 pounds. So if there are 10 50-lb boxes in a stack on the floor, the solvent of people can pick up all 10 boxes, leading to complete dissolution of the box pile in the people solvent. However, if 10 250-lb boxes are on the floor, five people are required to pick up one box, meaning the group can only pick up two boxes from the pile. Thus, part of the box stack solute is dissolved in the people solvent, but the rest of the box stack is undissolved. While the types of forces and chemical details are more complicated in chemical solubility than the box example, the principle is the same: solvents often have a limited capacity to carry solutes, and if you place more solute particles in solution than the solution can support, a solid will form. This clearly can be a problem in your car. You would not want to precipitate a solid in the cooling system plumbing, since that would clog pipes and prevent a flow of coolant, perhaps leading to engine overheating and serious damage (see Chapter 5 for more info on engine heat management). Likewise, you do not want a significant amount of water to dissolve in your ethanol-enriched gasoline, as water can damage the engine, and too much water hinders the combustion process.

Aqueous Fuel Emulsions

Several patents in the 1960s and 1970s relate to the development of aqueous fuel emulsions for use in primarily diesel-based combustion engines. These emulsions offer several benefits. For example, accidental diesel fires in emulsion fuels are automatically self-extinguished. In addition, the presence of significant water in the fuel causes the combustion temperature to be reduced, which lowers the formation of soot particles and causes the temperature of the exhaust gases to be lower, thereby limiting NO and NO_2 gas produced as combustion by-products (the latter gases are major contributors to smog). However, we know that diesel fuel is primarily nonpolar organic molecules, while water is a highly polar solvent, meaning that the two should not mix. Thus, to develop an aqueous emulsion suitable for use as a fuel, we must overcome the inherent lack of solubility of hydrocarbons in water using chemistry. This is done using surfactant molecules similar to those that make up automotive detergents and discussed in Section 4.3. Surfactants are often large molecules that have regions that are dominated by dispersion forces that are soluble in hydrocarbons and regions that can readily interact with polar solvents such as H_2O. Many of these surfactant additives for fuel emulsions involve fatty acids such as oleic acid ($C_{18}H_{34}O_2$) and amine/amide compounds such as diethanolamine ($C_4H_{11}NO_2$) and diethanolamides. However, generating a stable emulsion can be quite complicated, and many of the available patents list relatively complex chemistries involving surfactants, hydrocarbon fuels, water, antifreeze agents such as ethanol to prevent freezing of the emulsion, and other specialty chemicals to provide storage or combustion advantages.[*][†] The selection of appropriate chemicals is crucial to developing a viable emulsion fuel, and this selection is rooted in the topics of intermolecular forces and solubility.

---

[*] G. E. Fodor, W. D. Weatherford, and B. R. Wright, "Fire-Safe Hydrocarbon Fuels," US Patent 4,173,455 A, filed Oct. 11, 1978, and issued Nov. 6, 1979.

[†] D. T. Daly, J. J. Mullay, E. A. Schiferl, D. A. Langer, D. L. Westfall, H. Dave, B. B. Filippini, and W. D. Abraham, "Emulsified Water-Blended Fuel Compositions, US Patent 6,280,485 B1, filed Sep. 7, 1999, issued Aug. 28, 2001.

4.3   Detergents

Chemistry Concepts: intermolecular forces, organic chemistry, hydrophobicity/hydrophilicity

Expected Learning Outcomes:

- Define surfactant
- Describe the structures of an organic molecule that would make a good detergent
- Define the terms *hydrophobic* and *hydrophilic* and relate them back to intermolecular forces
- Describe how a surfactant acts as a detergent

Keeping an automobile clean is important aesthetically and is essential to the longevity of a vehicle, particular in locations where the car is exposed to harsh chemical conditions that accelerate corrosion. As we discussed in the previous chapter, chemicals such as road salts, salty aerosols near oceans, and acid rain all promote iron/steel corrosion and may damage plastic components or even the pigments in paint. The surface of your car is exposed to dust, tar, oil, pollen, and other road grime that traps chemicals against the paint surface or may scratch the protective coatings. It is important to remove these contaminants, yet it is equally important not to damage the paint, strip the protective wax coating, or harm the many metals, alloys, and other structural or aesthetic materials in the process. There are an enormous number of different car-cleaning agents on the market, from soaps to soapless washes to specialty cleaners like engine degreasers and spray-on wheel cleaners. Many of these cleaning agents, particularly the general-purpose soaps, contain or may contain dyes, brighteners, and waxes to promote shine; ions to maintain charge neutrality; alkalis to modify the pH; and in all cases some type of surfactant molecule. Of all these components, it is the surfactants that actually do a majority of the cleaning. In this section, we discuss the remarkable physics and chemistry of surfactant automobile detergents and stress the role of solubility and intermolecular forces in their function.

A surfactant is any chemical that alters the surface tension of a liquid at its interfaces with itself, other liquids, or gases. Surface tension is the force that develops at the surface of a liquid due to the stronger intermolecular interactions between the molecules at the surface and those just below the surface compared to liquid molecules in the bulk. Only half of the surface molecule is capable of making contact with other liquid molecules at any given time, meaning it is pulled downward toward the liquid. Surfactants for use in water-based systems are nearly always long-chain polar organic molecules that have both hydrophilic (water-loving) and hydrophobic (water-hating) regions. Hydrophilicity is the degree to which a substance forms strong intermolecular interactions with water. Hydrophilic chemicals are those that can participate in hydrogen bonding, ion–dipole, and/or dipole–dipole interactions and thus interact strongly with water. Hydrophobic compounds are those that do not easily develop these types of inter-molecular forces, meaning those compounds that are dominated by dispersion-type interactions. In surfactants for use with water, the long hydrocarbon chain is the hydrophobic component and is terminated on one end with some type of hydrophilic group that allows the molecule to be water soluble.

There are many types of hydrophilic terminal groups used in sur-factants; indeed, surfactants are generally categorized according to the nature of the hydrophilic group terminating the hydrocarbon chain. The three general categories include anionic surfactants, cationic surfac-tants, and nonionic surfactants. Anionic surfactants are those that have an anion at the hydrophilic end of the molecule. The most common anions in anionic surfactants include sulfate, sulfonate, phosphate, or carboxyl-ate groups. However, detergents containing phosphate are now banned due to environmental concerns. Cationic surfactants are terminated with cationic hydrophilic groups such as amines or ammonium salts. There is also a special category of ionic surfactant known as a zwitterionic surfactant that contains both positive and negative ionic groups at the hydrophilic terminus; additional discussion of zwitterions is beyond the intended scope of this book. The nonionic surfactants do not use charged species at the hydrophilic terminus and are most commonly long-chain alcohols that are weakly polar.

It is precisely this amphiphilic (mixed hydrophobic and hydrophilic) property of surfactants that make them excellent automobile detergents. The solvent for washing a vehicle is nearly always water, as it is cheap, abundant, and capable of dissolving many different types of materials. However, water is a very poor solvent for oils, tars, and other common contaminants on the road that are nonpolar, hydrophobic materials. To remove the hydrophobic organics from your car, we need the proper-ties of long-chain hydrocarbons in a water-soluble form—a surfactant. Although the surfactant is soluble, the long hydrocarbon end does not wish to be in water and will selectively interact with hydrophobic con-taminants on the vehicle surface (Figure 4.2). With the hydrocarbon end of the molecule "dissolved" in the hydrophobic contaminants, the hydro-philic end faces outward toward the water. Let's consider an oil droplet as a typical hydrophobic contaminant on your car. Once enough surfactant molecules have attached their hydrocarbon ends to the oil droplet, it appears to be a soluble material from the perspective of the water, which can only encounter the hydrophilic termini with which it can easily inter-act. The water is now capable of lifting the oil droplet from the surface and carrying it off of the car.

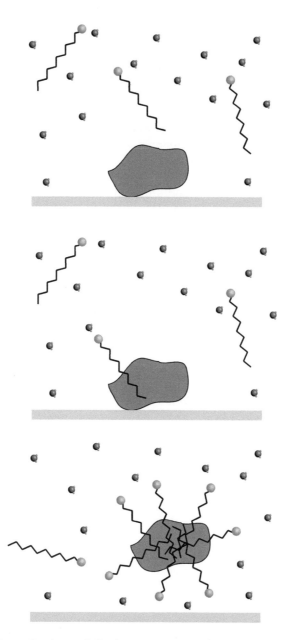

Figure 4.2 (See color insert.) Surfactant action with an organic contaminant. In the top image, we see the water far away from the oily droplet and surfactant molecules in solution. In the middle image, surfactants begin to dissolve their hydrophobic ends in the oil droplet, bringing some water closer to the dirt. In the final image, the droplet is coated in surfactant molecules, making it look water soluble and allowing it to lift from the surface into the water solvent.

Most commercial detergent-based automotive soaps contain either sodium lauryl ether sulfate or dodecylbenzene sulfonic acid (Figure 4.3) as their surfactant. These are both anionic surfactants, and in both cases, the chain terminates with an oxygen anion that can participate in ion–dipole interactions and hydrogen bonding, making it highly soluble in water. Although these molecules are also polar molecules, the polarity is relatively low and is not a major contributing factor to their solubility in water. Some cleaning agents avoid the use of these surfactants. For example, waterless/soapless car washes are available that use simply oils, lubricants, and proprietary chemicals to encapsulate and lift both inorganic and organic contaminants from the car surface without the use of water. Citrus-based cleaners are now relatively common in the home-care industry and are starting to appear in the automotive cleaning market. High-quality citrus cleaners are based on citrus oils, which can effectively dissolve a variety of organic compounds. The active ingredient in most of these cleaners is d-limonene (Figure 4.4),

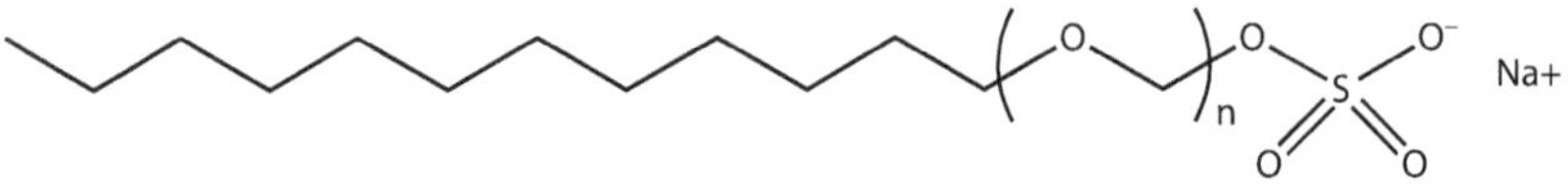

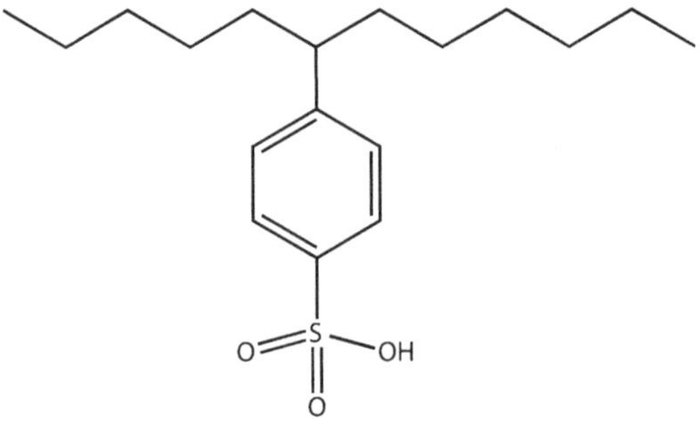

Figure 4.3 Chemical structures of sodium lauryl ether sulfate (top) and dodecylbenzene sulfonic acid (bottom).

Figure 4.4 Chemical structure of d-limonene, a common terpene found in citrus-based cleaning agents.

which is a terpene compound commonly produced by plants. One must be careful about water-based citrus cleaners, however, as the citric acid found in these cleaning agents may actually promote corrosion by providing a H⁺ source. There are positive and negative aspects to all types of cleaners discussed in this paragraph, and the right choice for you is likely to be more a matter of personal values than having one clear-cut option that is better for your vehicle. The important thing in car care is to clean your car of contaminants often and reapply the wax coating regularly, since most cleaners slowly remove this protective layer from the vehicle surfaces.

4.4 Lubricants

Chemistry Concepts: properties of liquids, organic chemistry, intermolecular forces

Expected Learning Outcomes:
- Understand the concept of viscosity and how it varies with temperature
- Explain how to alter oil viscosity on a chemical basis
- Describe the various roles played by transmission fluid
- Understand how grease differs from motor oil

There are many places in a car where two solid surfaces are in contact, providing the potential for significant heating and wear due to friction. To manage the consequences of friction in critical components such as the engine and transmission, a variety of lubricating agents are used in modern vehicles. These lubricating agents both reduce the coefficient of friction between solid contacts and carry away waste heat as they are circulated, helping to keep components cool. At the same time, minor additives clean engine parts, provide corrosion protection, and stabilize the lubricants against chemical degradation. In this section, we discuss several types of lubricating agents in cars, their common properties, and how these properties relate to intermolecular forces.

Perhaps the most important quality of lubricating agents is viscosity. Viscosity is the ability of a fluid to resist flow and arises largely from the intermolecular forces between lubricant molecules. For example, when pouring honey from a jar, the force of gravity on the honey is opposed by the attractive intermolecular forces between the molecules within the honey such as sugars and water. The intermolecular forces in honey are quite strong compared to the thermal motion of the molecules, leading to a thick, slow-pouring fluid that we would classify as having a high viscosity. On the opposite end of the spectrum, ethanol has a very low viscosity and flows quite freely at room temperature when poured from a container. It is also essential in your conceptual picture of viscosity to understand that fluid viscosity is a function of temperature. Honey placed in the refrigerator or freezer flows visibly more slowly than honey at room temperature. The change in viscosity as a function of temperature relates back to ideas we discussed in Chapter 1 (The Properties and Behavior of Gases)—higher temperature leads to greater thermal motion of the molecules. In fact, both the thermal motion and gravity oppose the inter-molecular attractions when pouring a fluid. In cold fluids, gravity has to do more work to overcome the intermolecular attractions, since there is less "help" from thermal motion; thus, viscosity always increases with decreasing temperature. This helps to explain how ethanol and water, both with very strong hydrogen-bonding interactions between mole-cules, flow more easily than honey: they are much smaller molecules and therefore have much higher mean molecular velocities at room temper-ature. The temperature dependence of viscosity is the reason that your car's manual likely suggests different weights of motor oil depending on the atmospheric temperatures where your car will operate, and this is the main driving force behind the use of dual-grade motor oils today.

Why is the viscosity of a lubricant so important? For one thing, the effectiveness of the lubricant pumping system is highly dependent on the viscosity. If a pump designed to move thin liquid with a low viscosity is fed a thick, viscous liquid, it likely will not generate the appropriate flow rate, starving critical components of lubricant and leading to vehicular damage. But aside from this mechanical concern, we need a lubricant viscosity that balances the two main engineering requirements of the lubricant: For a lubricating agent to reduce friction, it must cling to the solid surfaces where they come into contact, and the molecules in the lubricant must slide smoothly along one another. A thick, viscous lubri-cant will easily cling to metal surfaces; however, the strong intermolecular attractions will likely not allow the lubricant molecules to slide smoothly along one another. None of us would want a thick grease lining the cyl-inder walls in our engine. A thin lubricant contains molecules that slide easily along one another, but may not cling well to the surfaces of interest. For these reasons, oils of very specific viscosity are required in different components of your car, depending upon the role of that component, and we can use what we know about intermolecular forces to design an appro-priate chemical system for each situation. Beyond just the appropriate balance of adhesion and slip under operating conditions, the lubricating agent must also retain its lubricating properties at cold temperatures as well as resist degradation via chemical reactions under the hot oxidizing environments found in some car parts such as the engine under normal operating conditions. Many different types of additives—friction modi-fiers, stabilizers (antioxidants), corrosion inhibitors, and detergents—are added to motor oils to minimize the reactivity and thermal variations in viscosity, as noted in the opening paragraph for this section.

Your vehicle's engine is lubricated with a mixture of organic compounds that we call *motor oil*. Most motor oils are composed almost entirely of hydrocarbons with a small amount of additives, with the dominant hydrocarbons typically containing between 18 and 34 carbon atoms per molecule. The base oil, or the main component to which minor amounts of additive chemicals are added, is obtained either from petroleum distillation or synthetic routes such as the Fischer-Tropsch process. Fischer-Tropsch reactions convert hydrogen and carbon monoxide feedstocks into alkanes according to the following type of overall reaction:

$$(2n + 1)\, H_2 + nCO \rightarrow C_nH_{(2n+2)} + nH_2O$$

Recall that organic hydrocarbons are dominated by dispersion interactions and that dispersion interactions depend on the molecular weight of the molecules and their geometries. Therefore, by changing the molecular weights, conformations, and aspect ratio (linear versus. highly branched) of the chemicals in a motor oil, we exert control over the oil viscosity. Unfortunately, the actual chemical compositions of motor oils are often proprietary and highly protected, so the Society of Automotive Engineers (SAE) has developed performance criteria for rating motor oils based on their observed viscosities and other properties.[*] Motor oils for automobile engines are almost entirely SAE dual-grade oils that contain modifiers such that the viscosity at low temperatures is similar to the base oil and the viscosity at the high temperatures is similar to the modifier, allowing the same motor oil to be used throughout the entire year. The SAE weighting system on oils (e.g., 10W-30) indicates what SAE viscosity-grade tests the motor oil has passed. The presence of two numbers in our example indicates that this is a dual-grade oil. The "W" after the 10 indicates that this oil passes the Society of Automotive Engineers viscosity-grade-10 test at winter operating conditions. The 30 indicates that the oil passes the viscosity-grade-30 test at hot operating conditions. For more information on the testing procedures and viscosity grades, see the SAE J300 standards document. However, keep in mind that performance in these tests is ultimately determined by the types and strengths of intermolecular forces, which in turn depend on the weight and structure of the molecules in the motor oil.

Just like in the engine, the transmission contains lubricants that manage friction and temperature, prevent corrosion, and perform other important roles such as cleaning. The type of lubricant needed in your transmission depends on the type of transmission in your vehicle. Most older manual transmissions use gear oil, which is a nonpumped oil of relatively high

[*] "Engine Oil Viscosity Classification," Society of Automotive Engineers, Doc. J300_201304, 2013.

viscosity that can be either single grade or dual grade, while some more recent manual transmission designs make use of conventional automatic transmission fluids. Typically, gear oils form a thin layer in the bottom of a pan, and the rotating gears themselves make contact with the oil, thereby dispersing the oil in the transmission. Gear oils are generally of a higher viscosity than motor oils, but they are also prepared by incorporating chemical additives into a natural petroleum-based oil or a synthetic base oil. The base oils for gear oil contain chemical compounds such as oligomers (polymers with only a few monomer units per polymer chain) made from hydrocarbon alkenes with one double bond located at one end of the hydrocarbon chain (alpha-olefins), diesters, polyol esters (esters containing several –OH groups), alkylbenzenes, and alkylnaphthalenes with carbon contents of 12 to 120 carbon atoms per molecule. These lubricants are also dominated by dispersion interactions and, since they involve larger molecules, develop larger dispersion interactions than a typical motor oil—thus the higher viscosity.

Automatic transmission fluids (ATFs) differ slightly from typical gear oils and often have more complicated chemistries. Base oils for automatic transmission fluids are typically saturated hydrocarbons (up to 98%) with between 10%–40% cyclic organic compounds. As with gear oils, they have relatively high average molecular weights compared to typical motor oils, with nearly 50% of the molecules containing 30 or more carbon atoms. The base oils are once again obtained from petroleum distillation or prepared via a variety of chemical reactions, including the Fischer-Tropsch process, hydroisomerization reactions, and other cracking processes combined with distillation methods for separating the various oil fractions. Hydroisomerization reactions convert large *n*-alkane hydrocarbons into branched alkanes, which will reduce the dispersion forces while promoting entanglement of the chains, and cracking reactions convert large alkanes into smaller alkanes, reducing the dispersion forces. Automatic transmission fluids also contain modifiers to maintain band friction and prevent corrosion as well as dyes to identify the fluid if it is leaking. The interested reader can learn much more about the composition and preparatory chemistries for these lubricants by delving into the patent literature.[*][†]

Thus far, we have focused on "wet" lubricants, but cars also contain a variety of dry components where friction management is required, such as bushings and suspension components, wheel bearings, brake-caliper slide pins, etc. These dry environments require much more viscous and long-lasting lubricants known as greases. A grease is generally prepared by mixing an oil with a surfactant thickening agent. This creates an emulsion,

[*] S. Hara and T. Okada, "Gear Oil Composition," US Patent 20,100,113,314 A1, filed Mar. 19, 2008, and issued May 6, 2010.
[†] G. R. B. Germain, H. D. Mueller, and D. J. Wedlock, "Automatic Transmission Fluid," World Patent 2,002,070,636 A1, filed Mar. 5, 2001, and issued Sept. 12, 2002.

which is a mixture of two immiscible liquids that are dispersed evenly in one another. You can see an example of an emulsion by shaking up an oil-and-vinegar salad dressing. The whitish creamy liquid that results is an emulsion of the oil in the vinegar, and the viscosity of the emulsion is often higher than either of the two pure components. The viscosity of the greases used in automotive applications is so high that the emulsion is considered to be a pseudo-plastic fluid, or a fluid whose viscosity is reduced under shearing forces. Shearing forces occur when one part of an object pushes in one direction and another part in the opposite direction, much like a pair of scissors cutting paper. This behavior allows the grease to stick to the components under most conditions, yet provide lubrication by sliding past itself easily when the component it is coating is under stress. The advantage of grease for dry applications is that it does not need to be replaced often. Motor oil needs to be changed every 6 months or 3000 miles (though some modern normally aspirated engines now can go 7000–10,000 miles between oil changes), in contrast to lubricants in hard-to-access wheel bearings or suspension bushings, which do not need to be changed so often.

4.5   Wax

Chemistry Concepts: intermolecular forces, organic chemistry, hydrophilicity

Expected Learning Outcomes:
- Explain how wax protects a car
- Describe the molecular characteristics of a good wax
- List several types of waxes used in commercial automotive waxes

In the previous chapter, we discussed many methods of preventing corrosion in automobiles. Aside from the zinc galvanization, all of the corrosion-prevention methods for the vehicle body focused on keeping water and ions from reaching the steel. As a car owner, you have little control over the electrocoat, base coat, paint, or clear coat on your vehicle; repainting is always an option, but often a very expensive one. However, you can control the outermost layer of corrosion protection—the wax. Waxes protect the body panels by preventing wetting, which is a sophisticated way to say that they repel water. How do they repel water? Waxes are all very hydrophobic materials. A good wax will not be able to participate in hydrogen bonding, dipole–dipole interactions, etc., and thus should contain primarily long-chain organic molecules with a low oxygen content. By applying a thin coating of these materials to the surface of your paint, they force the water and any corrosion-promoting chemicals it contains to bead up on the surface and easily run off the body as large, heavy droplets. The wax coating does wear away with time and is susceptible to degradation via heat or UV radiation (see Chapter 7) and to the cleaning action of detergent surfactants. High temperatures may cause the more volatile components of the wax to vaporize and leave the surface. Many types of ultraviolet radiation have an appropriate amount of energy to induce photohydrolysis reactions of esters or hydration reactions across C=C, changing the molecular structure and ultimately reducing the mean molecular weight of the wax molecules and hydrophobicity of the wax. However, applying a wax to your car several times a year goes a long way to protecting your vehicle, especially when the outer protective coatings are scratched. In addition, waxes provide a rich, glossy finish to your vehicle that is aesthetically pleasing and often helps to filter out UV radiation, preventing photochemical reactions of the paint pigments.

There are several different types of automotive wax, but they all share a number of chemical similarities, such as the carbon chain lengths and dominance of dispersion interactions between molecules. Natural waxes are often the most desirable and are those derived from plants and other renewable resources, such as carnauba wax. Carnauba wax provides a great deal of shine and offers excellent water protection. It is derived from a type of a Brazilian palm tree (*Copernicia cerifera*) that grows only in a specific region of the country. Chemically, carnauba wax contains a series of organic compounds that contain 24–34 carbon atoms per molecule, including aliphatic esters, alpha-hydroxy esters, and cinnamic aliphatic diesters as well as free acids, free alcohols, hydrocarbons, and resins (thick natural liquids that harden into mostly transparent solids). It tends to have a sweet, bananalike odor and is sometimes also used in food products. The main drawbacks to carnauba and other natural waxes are that they last only short periods of time before you need to recoat the car (less than two months), and it is difficult to remove the excess wax during the application process. To overcome these limitations, companies have explored a variety of alternative synthetic waxes. The most common synthetic waxes are paraffin waxes and montan waxes. Paraffin waxes are either oil-derived or synthetic hydrocarbon waxes in which each hydrocarbon chain contains ≈20–40 carbon atoms. Montan waxes are produced by extracting compounds from lignite (brown coal) using naphthalene or benzol. These waxes also contain long-chain organic acids, alcohols, and ketones with mean carbon-chain lengths of 24–30 carbon atoms. Montan waxes and paraffin waxes are both longer lasting than natural waxes like carnauba; however, they tend not to produce the same type of shine. There are also synthetic silicone waxes available, which are based on organo-terminated polysiloxane chains rather than organic carbon-based chains, for example, polydimethyl siloxane (Figure 4.5). By having a siloxane core with methyl and other organic functional groups at the chain exterior, one can again make a dispersion-dominated molecule that will be highly hydrophobic. Silicone waxes are also easier to remove, as silicones and

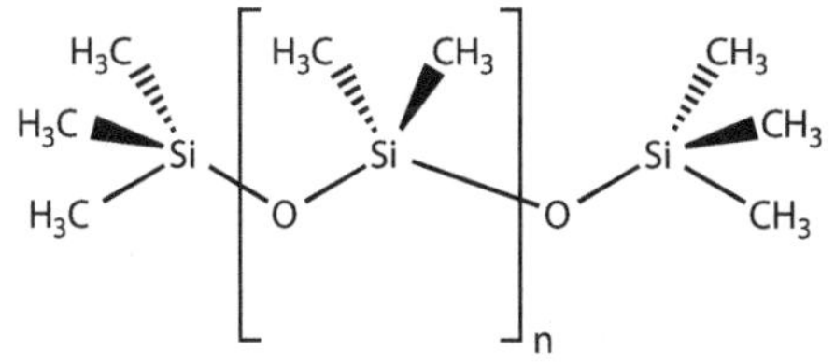

Figure 4.5 Chemical structure of polydimethyl siloxane, a typical organofunctionalized silicone polymer.

silicone oils make good lubricants. Silicones are also popular additives to help stabilize wax emulsions, enhance the shine, and make it easier to remove excess wax during buffing. Many commercial waxes also contain solvents and detergents as well as fine abrasives to help clean the surface.

As noted, waxes protect the paint by providing a hydrophobic coating that forces the water to form large, heavy droplets on the surface. When water falls onto the dispersion-dominated system set up by the wax coating, the small polar water molecules develop only very weak dispersion interactions with the wax surface as a result of their small size. There are large numbers of water molecules in a typical raindrop, and all of these water molecules interact more strongly with each other via hydrogen bonding and dipole–dipole interactions than they do with the wax coating. Thus, rather than wetting the surface, the water molecules pull one another away from the surface to form tall droplets that have a more spherical shape. As additional water strikes the surface, it will be drawn into droplets that are already on the surface rather than wetting additional areas. As the drops become large and heavy, the forces of gravity and friction between the droplet and the air flowing around the vehicle exceed the weak attractive forces between the water and wax film, causing the droplet to run off the surface and preventing water from coming into contact with the metal and the more-permanent protective coatings. Intermolecular forces are therefore critical to the protective action of a wax coating.

There is confusion about the role of waxes and polishes for vehicles, which is understandable given their many chemical similarities. The primary role of a polish is to clean the vehicle surface during detailing, and polishes contain abrasive powders that physically remove a very thin layer of the paint finish. In addition to the powdered abrasive, polishes often contain silicones in the form of polydimethyl siloxanes, aminofunctional silicones, and silicone resins to help spread the powder as well as provide shine and some protection for the paint surface. Polishes may also contain some amount of natural or synthetic wax, solvents, emulsifiers (often silicones), and thickeners such as clays or gums. Some polishes also contain dyes and water-resistant resins. In general, polishes do a better job of cleaning the paint surface, and waxes do a better job of protecting the paint.

chapter five

Managing heat

Energy losses to heat are the heart of inefficiency in automobiles. In some cases, this heat loss is intentional (conventional braking systems), and in others it is a practical cost of current component designs (the combustion engine). We have discussed several mechanisms in cars to recover wasted heat energy thus far, but none of these methods can completely manage the heat produced by critical components of the car. Overheating has numerous drawbacks, with possible consequences including minor warping and other forms of deformation in components to outright structural failures. Passenger comfort is also a significant issue for manufacturers to address in the modern car market, and that means managing the temperature of the cabin. Air conditioning systems currently appear as standard equipment on nearly all modern automobiles and are essentially heat-exchange devices. Thus, managing heat through heat exchangers, refrigeration systems, and well-selected materials or component design is critical to the function and durability of cars. In turn, heat exchange and management rely in many aspects upon chemistry. In this chapter, we discuss several fundamental chemical aspects of heat-transfer equipment and the function of heat-exchanging devices from a chemical perspective.

5.1 Colligative properties and your antifreeze

Chemistry Concepts: solution chemistry, thermodynamics, intermolecular forces

Expected Learning Outcomes:
- Understand the molecular basis for boiling-point elevation and freezing-point depression
- Explain why including ethylene glycol in antifreeze is beneficial
- Understand how road salt helps prevent ice formation
- Understand why water itself is both a good and a poor radiator fluid

Most vehicles manage the waste heat produced by the engine via an air-to-liquid cooling system, though some vehicles can still be found on the road that are purely air cooled, such as late-model Porsche 911s. In a liquid-cooled engine, a pump circulates a liquid through the engine block and other critical engine components, then through a device called a *heat exchanger* external to the engine. The liquid in the cooling system picks up thermal energy in the form of heat spontaneously as it passes through the engine, then gives up this heat energy to the atmosphere spontaneously as it passes through the heat exchanger. The liquid engine coolant must be capable of withstanding a wide range of temperatures and avoid changing phase under normal operating conditions. We have already discussed that temperatures in the engine combustion chamber and exhaust systems can be very high. At the same time, the coolant must also be able to remain a liquid well below the freezing point of water in cold climates. Despite the seemingly harsh conditions at opposite ends of the engine heat-management spectrum, water remains a significant chemical component in coolant, primarily because it is abundant, inexpensive, and has a very high heat capacity. *Heat capacity* is the ability of a material to store energy in the molecular-scale motions of the atoms and molecules. But how can we use water-based coolants and maintain the liquid state over the full range of possible operating temperatures? One reason is that the cooling system is under pressure, and the higher pressure shifts the boiling point to well above 100°C. But a second and perhaps more important reason is that engine coolant is actually a solution that takes advantage of the colligative properties of all solutions.

Colligative properties are solution properties that depend only upon the ratio of solute particles to solvent particles, not on the identity of the solute. Most general chemistry texts mention several important colligative properties in the context of aqueous solutions like our modern engine coolant, including boiling-point elevation and freezing-point depression. Boiling-point elevation means that a solution will always exhibit a higher boiling point than the pure solvent. Often, we take advantage of this colligative property when cooking pasta. Adding a small amount of salt to the water increases the boiling point slightly (fractions of a degree) and helps to cook the pasta faster. A solution also resists freezing more intensely than the pure solvent, causing the solution to freeze at a lower temperature (freezing-point depression). You can observe this in your kitchen as well if you put a bottle of water and a bottle of vodka in your freezer. The typical vodka is a 40% ethanol and 60% water solution and will not freeze, while the water bottle should freeze completely. Other important colligative properties, such as osmotic pressure (a complete discussion of which is beyond the scope of this text), are usually presented in more advanced courses in physical chemistry. From the perspective of automobile coolant chemistry, it is boiling-point elevation and freezing-point depression that matter the most.

On a molecular scale, we can use intermolecular forces and the concept of physical equilibrium to explain the origin of boiling-point elevation and freezing-point depression. A liquid boils when a temperature is reached such that the liquid and vapor states of a material are in

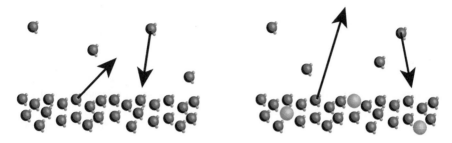

Figure 5.1 (See color insert.) The pure solvent (left) and solution (right) liquid–vapor interfaces. The length of the arrows depicts the energy needed to change phases and shows that the solute upsets the balance between the energies, meaning that the temperature must be increased to reestablish equilibrium.

dynamic equilibrium at the system pressure, which means that the rate of evaporation is identical to the rate of condensation. To understand boiling-point elevation, picture the interface between pure water and steam (Figure 5.1). To exit the liquid phase and join the vapor, a water molecule must acquire enough kinetic energy to overcome all the inter-molecular forces and leap off of the liquid surface into the vapor. For a steam molecule to enter the liquid state, it must collide with other molecules in the steam or solvent molecules at the liquid surface with a low enough kinetic energy such that the intermolecular forces grab and hold on to the H_2O molecule from the steam, causing it to stick. In the pure solvent case, the interface between liquid and vapor is made up entirely of solvent molecules. For aqueous solutions, it is known that the vapor phase remains ≈100% steam. In other words, it is safe to assume that the solute only exists in the liquid phase. Thus, over a solution, the rate at which a steam molecule condenses to liquid remains essentially the same as in the pure solvent case at any temperature. Without any solute around, the rate at which a solvent molecule in the vapor strikes another vapor-phase solvent molecule with low enough kinetic energy to stick and form a droplet remains constant, since the vapor phase is still pure steam. However, since the solvent now contains solute particles, the activity of the solvent is reduced compared to the pure solvent state because of the presence of solute. On a molecular scale, it is likely that some solute particles are at the interface at all points in time, meaning that the rate of a water molecule in the liquid striking the liquid–vapor interface with high kinetic energy and leaping into the vapor phase is reduced compared to the pure solvent state, where the interface is made up entirely of solvent. This removes the equilibrium condition at the pure solvent boiling point, because now the attachment and departure rates of solvent molecules are not identical. To reestablish the equilibrium condition by making the condensation rate and evaporation rate identical, more heat must be added to the liquid phase, with the net result being an increase in the boiling point. Similarly, a liquid freezes when the rate of attachment of liquid molecules to the solid phase is equal to the rate of detachment of a solid phase molecule into the liquid. Solutes are generally insoluble in ice, leading to a similar situation to that of the boiling liquid described here: The presence of the solute reduces the solvent activity by diluting the solvent in the liquid phase, thereby reducing the liquid attachment rate while not affecting the particle detachment rate from the ice. Thus, an additional reduction in temperature is required to increase the liquid-to-solid attachment rate and reestablish the equilibrium criterion, leading to a reduction in the freezing point.

The coolant solution in your automobile engine is predominantly a mixture of water and ethylene glycol (Figure 5.2) along with chemical additives to prevent corrosion (such as zinc dialtyldothiophosphates [ZDDP], discussed in Chapter 3), dyes, etc. As noted in the previous chapter, to effectively dissolve a substance, the solvent and the desired solute must be capable of generating similar intermolecular forces. In water, hydrogen bonding is the dominant intermolecular interaction. Looking at the structure of ethylene glycol (EG), one can see that two hydroxyl (−OH) functional groups are present. Both of these hydroxyl groups can actively take part in a hydrogen-bonding network by donating and receiving hydrogen bonds, meaning that EG has a relatively high solubility in water. Typical engine coolant is up to 50% water, 50% EG, making it a very concentrated solution rather than the ideal dilute solutions we usually consider in general chemistry. This means we should work with activity rather than concentration to predict the true behavior. However, we can understand the influence of EG on the thermal stability of water using the relatively poor assumptions that concentration and activity are equivalent in a typical engine coolant and by treating the solution as ideal.

To find the boiling point of our coolant solution, we must know the boiling point of water at the pressure of the cooling system and the colligative boiling-point elevation provided by the presence of the EG. In the case of water, many reference books contain tables and plots of the boiling point as a function of pressure. If you look at the radiator cap on your car, you should find a pressure rating. Most cooling systems are kept under 15 psi gauge, so we will need to use the total coolant pressure of ≈30 psi or ≈2 atm to identify the correct boiling point for water in the cooling system.

Figure 5.2 The chemical structures of water (left) and ethylene glycol (right).

The boiling point of water at this pressure is 121°C. Clearly, pressurizing the cooling system has a significant influence on the upper temperature limit of the liquid coolant's thermal stability zone. We can also easily calculate the colligative influence of the EG on the boiling point using the following equation:

$$\Delta T = iK_b m$$

where i is the Vant Hoff factor (which has a value of 1 for a solute that does not dissociate, such as ethylene glycol), K_b is the boiling-point elevation constant of the solvent (0.512 for water), and m is the molality of the solution in moles of solute per kilogram of solvent. For a 50/50 (by mass) mixture of ethylene glycol and water, we can determine the molality and boiling-point elevation using a 100-g basis:

$$50 \text{ g EG} \times \frac{1 \text{ mol EG}}{62.07 \text{ g}} = 0.805 \text{ mol EG}$$

$$\frac{0.805 \text{ mol EG}}{0.05 \text{ kg solvent}} = 16.11 \text{ molal}$$

$$\Delta T = 1 \times 0.512 \, \frac{°C}{\text{molal}} \times 16.11 \text{ molal} = 8.24°C$$

Thus, the boiling point of our engine coolant under 15-psi additional pressure is ≈129°C. The calculations also show that the boiling-point elevation due to pressurization is more important than the boiling-point elevation due to the colligative properties of engine coolant. This exercise is also clear evidence that you should regularly monitor whether your radiator cap is maintaining the proper overpressure. If the pressure drops, it makes it much easier for the coolant to vaporize and dramatically increases the chances of your car overheating.

A similar analysis can be used to calculate the effect of the ethylene glycol on the freezing point of the coolant. The freezing point of water is much less variant with pressure than the boiling point, so much so that we can consider the freezing point of water at 2 atm of pressure in the cooling system to be effectively 0°C. To determine the colligative freezing-point depression, we can use the same equation as for boiling-point elevation, but we must replace the boiling-point elevation constant with the cryoscopic constant K_f, which for water is 1.85 K molal^{-1}:

$$\Delta T = 1 \times 1.85 \, \frac{°C}{\text{molal}} \times 16.11 \text{ molal} = 29.8°C$$

Thus, engine coolant will not freeze until the coolant temperature reaches the natural freezing point minus this change, giving the solution a freezing point of ≈−30°C.

Keep in mind that since the boiling-point elevation and freezing-point depression are colligative properties, it does not matter what solute we use as long as it is soluble in water. A 16.11 molal solution of ethanol/water mixture will provide the same boiling-point elevation and freezing-point depression, as well as meeting the solubility criterion. However, this mixture may be flammable in an accident and may induce more serious corrosion problems in the cooling lines. Thus, we must consider both solubility and practical issues such as safety and cost when selecting a good solute for an effective engine cooling system that helps your engine have a long lifetime.

Road Salt

Colligative properties of solutions are also used to keep roadways safe during the winter in cold climates. Ice and snow make it difficult for a vehicle's tires to make contact with the road surface and generate the friction required to move the vehicle. Snow plows can remove a significant portion of the snow, but mechanical removal methods cannot efficiently deal with ice on the roadway. However, we now know that if we can generate an aqueous salt solution above the ice, the freezing point of that solution will decrease, and we can prevent ice formation. But how can we get an aqueous solution above the ice or melt existing ice when it is below freezing outside? The materials we use as road salt (predominantly calcium chloride or magnesium chloride) are hygroscopic, which means that they are "water-loving" materials. Once deposited on the road surface, the salt will start to take up H_2O molecules from the air. As it does so, ions near the surfaces of the salt crystals become hydrated. These hydration events are exothermic, releasing a small amount of heat every time an H_2O molecule enters the coordination shell of an ion. Much of this heat energy goes toward overcoming the lattice energy that holds the salt crystal together. However, the combination of the enthalpy release and the increase in entropy (distribution of energy among available states; see Section 3.2) of the aqueous solution versus pure solid ice causes the liquid salt solution to have a lower free energy than the pure ice or salt. Free energy is really a measure of the total entropy change of the universe in any process from the perspective of the system, and spontaneous processes are associated with negative free-energy changes (Section 3.2).

Thus, the free energy released during the hydration process melts the ice. Once the ice has melted completely, we are left with a salt solution that has a freezing point well below the air temperature thanks to freezing-point depression.

5.2 The radiator

Chemistry Concepts: heat, mass transport, heat capacity
Expected Learning Outcomes:
- Describe the design of an efficient radiator
- Understand how a radiator removes heat from an engine

In Section 5.1, we discussed the solution chemistry of engine coolant and mentioned a device known as a heat exchanger that facilitates the exchange of heat energy between the coolant and the atmosphere. The heat exchanger for engine coolant in your car is called a *radiator*. The radiator presents a physical barrier between the liquid coolant and the gaseous atmosphere, and thus serves as an intermediate between these two heat reservoirs. A heat reservoir is a large collection of molecules storing thermal energy. Thus, an effective radiator must accept the heat efficiently from the coolant and provide an efficient mechanism for the atmosphere to remove heat from the radiator materials. A radiator in a car also needs to be compact to fit in the vehicle, to be constructed of lightweight material with a high heat-transfer coefficient to avoid increasing vehicle weight significantly, and present a large surface area to the atmospheric gases, since gases are mostly empty space, as we learned in Chapter 1. Heat transfer in general is a broad topic in engineering and science, but in this section we simply focus on explaining the molecular-level details of a radiator and how it accomplishes its objective.

First, let us focus on passing heat from the engine coolant to the radiator itself. As the coolant gets hot, the molecules in the coolant will translate in higher energy states. There is also a small possibility that they will rotate or vibrate in higher energy states as well; however, translational motion is almost always the mechanism of molecular motion most responsible for heat capacity. If we restrict ourselves to consider just translational motion, the net result is that molecules in the hot coolant will collide with the walls of the radiator with a greater frequency and with greater kinetic energy than they would in cold coolant. When they do so, the collisions will be inelastic, and the coolant molecules will transfer some of their energy to the atoms/molecules that make up the radiator. Since the radiator is typically made of solid aluminum metal, this energy takes the form of more rapid atomic vibrations in the metal crystal structure. Analogous thought experiments can be applied to extra vibrational and rotational energy of the coolant as well. Thus, to remove as much heat from the coolant as possible, we want a large surface area for the liquid–radiator interface that maximizes the number of coolant-to-radiator collisions. Keeping in mind our engineering requirement of a compact device, we can maximize this contact area by bending a long length of tubing many times to fit a large surface area in a small volume (Figure 5.3).

The heat energy absorbed by the radiator material needs to be transferred to the atmosphere. Via essentially the same process by which the radiator removes energy from the liquid coolant, atmospheric gas molecules will collide with the external surface of the radiator materials. When they strike the radiator, they will absorb some of the vibrational energy in the crystalline lattice and translate (and to a much lesser extent rotate or vibrate) at a slightly greater rate, increasing the local air temperature and decreasing the temperature of the radiator material. Thus, to effectively remove heat from the radiator materials, we need to maximize the contact area between the atmospheric gases and the radiator as well as achieve a high flow rate of cooling air past the radiator. Since gases have much lower density than liquids, the surface area of the heat exchanger in contact with the atmosphere must be much higher than the surface area of the heat exchanger in contact with the liquid coolant to obtain an appropriate number of gas-to-radiator collisions. The easiest way to accomplish this design requirement is to cover the radiator tubes containing the liquid coolant with thin, closely spaced metal fins. Provided that the heat-transfer coefficient of the radiator material is high, as it will be for aluminum and most other common radiator materials, the heat picked up from the liquid coolant will be efficiently spread along all of the radiator fins. The cooling-air flow rate can be made quite high by

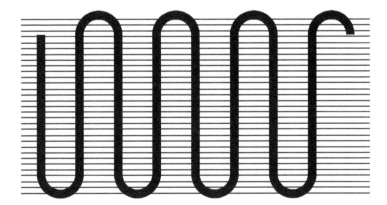

Figure 5.3 A typical radiator. The thick tubes represent the path of the liquid coolant and the thin lines the closely spaced fins on the radiator tubes.

placing the radiator near an opening in the vehicle facing the direction in which the car is moving. In commercial vehicles, this means placing the radiator immediately behind the grille and leaving an opening in the front of the car for the air to enter and pass by the radiator. This is also the reason why a car has radiator fans. If the car is not moving and a large amount of heat energy is still stored in the coolant, it may be necessary to force air through the radiator mechanically via the fans to prevent the coolant from overheating.

A more detailed discussion of the thermodynamics of heat transfer and heat exchanger design can be found in chemical and mechanical engineering textbooks.

5.3   *Refrigerants and atmospheric chemistry*

Chemistry Concepts: colligative properties, thermodynamics, gas laws
Expected Learning Outcomes:
- List common features of effective A/C refrigerants
- Explain the basic design of an automobile air conditioning system
- Explain the harmful role traditional A/C refrigerants have on the atmosphere
- Discuss why newer generation refrigerants have a lesser impact on atmospheric chemistry

Climate control of automobile passenger cabins dates to some of the earliest automobiles built by Packard in the late 1930s. However, widespread use of cabin cooling via air conditioning systems didn't occur until smaller and more efficient designs became available in the 1960s. Air conditioning is now standard on most automobiles, many of which feature advanced climate-control systems and filtration systems to remove pollen, allergens, and other particulate pollutants from cabin air. Thus, modern air conditioning systems cool the air in the passenger cabin, remove moisture from cabin air, and improve the overall air quality of the passenger cabin. While the basic design principles of vehicle air conditioners have been relatively unchanged for many years, the refrigerants themselves play multiple significant roles in chemical processes within the atmosphere and on vehicle safety, and thus they remain a point of contention.

Fundamentally, an automobile air conditioner is a refrigeration system for the cabin air and functions much like the food refrigerator in your home. The main challenge for air conditioning in an automobile and any other type of refrigeration is thermodynamics. One needs to remove heat from a cold body and transfer it to a warmer body, which is a nonspontaneous process. In the case of your car, this means removing heat from the cool air in the passenger cabin and depositing it in the warm atmosphere we have in the summer. The only way to accomplish such

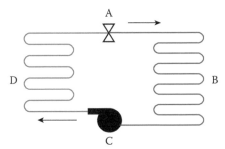

Figure 5.4 (See color insert.) Schematic diagram of the typical automotive air-conditioning refrigeration system. Just to the right/blue side of the evaporator valve (A), the coolant is a cold liquid at low pressure. It vaporizes by removing heat from the cabin air in the evaporator coils (B), then enters the compressor (C) as a cold vapor at low pressure. When it exits the compressor, it is a hot vapor that passes through the heat exchanger (D), giving off heat to the atmosphere and condensing into a hot liquid at high pressure.

a feat is to put energy into the system and take advantage of pressure/ temperature relationships and phase changes. Figure 5.4 shows a typical schematic diagram of the refrigeration system in an automotive air con-ditioner. If we start at the outlet of the condenser unit, the refrigerant is a warm liquid under substantial pressure. This liquid passes through an expansion valve, which drops the pressure and temperature in the cooling system so much that it is now possible for the liquid to evaporate with a small input of heat. The low-pressure liquid refrigerant passes through the evaporator unit, which is essentially a radiator that is in contact with a stream of flowing cabin air. As the cabin air passes over the evaporator unit, the liquid refrigerant removes heat from the cabin air and uses it to facilitate a phase change in the refrigerant from liquid to vapor. During this process, the cabin air is cooled and moisture is removed as it condenses on the cold evaporator coils. The low-pressure refrigerant vapor that exits the evaporator now passes through a com-pressor, which is where we put energy into the system: the compressor is turned by the engine and can consume as much as 2–4 hp. The com-pressor works just like a turbo or supercharger, boosting the pressure and temperature of the refrigerant vapor substantially. This hot, high-pressure vapor passes through a condenser that is in contact with the external atmosphere. The vapor temperature post-compressor must be warmer than the hottest expected atmospheric temperature to facilitate spontaneous heat transfer from the refrigerant vapor to the atmospheric gases. A good choice of refrigerant and boost pressure from the com-pressor make this possible. As the hot refrigerant vapor passes through the condenser, it gives up heat to the atmosphere and condenses into a high-pressure liquid. We are now back at our starting point, and one refrigeration cycle is complete.

A good refrigerant is any volatile chemical that can undergo a liquid-to-vapor phase transition at a low temperature under relatively low pressure (2–3 atm in a typical automotive system). A good automotive refrigerant is one that meets these criteria while remaining inexpensive, nonflammable, and environmentally friendly. Unfortunately, many of the early refrigerants were toxic and potentially dangerous—chemicals such as ammonia and methylene chloride that were quickly replaced following the discovery of chlorofluorocarbons, more commonly known as CFCs. The structures of the most commonly used CFC refrigerants in automobile air conditioning systems are presented in Figure 5.5. CFCs are easily produced by reacting chlorinated methanes and ethanes with hydrofluoric acid. They are nonflammable, unreactive, and have phase-change temperatures in ideal pressure ranges for the small, portable systems needed in cars. Unfortunately, it was later discovered that CFCs play important destructive roles in the stratospheric ozone cycle, leading to the ozone holes that develop over the poles in the polar spring. Essentially, these volatile chemicals are so unreactive in the troposphere, which is the region of the atmosphere where we live, that they can easily ascend without experiencing any chemical reactions. Once they reach the stratosphere, the higher intensity of UV-C light with sufficient energy to photolytically cleave one of the carbon–chlorine bonds dramatically increases the probability of producing a chlorine radical from the CFC:

$$CCl_3F + h\nu_{UV-C} \rightarrow CCl_2F\cdot + Cl\cdot$$

Radical species are very reactive, since they have unpaired electrons. The chlorine radical produced can react with ozone to generate perchlorate radicals and oxygen gas, reducing the amount of ozone present. With less ozone, more UV-B and UV-A light reaches the surface, which has negative consequences for plant and animal life. UV-B light is capable of cleaving bonds in DNA, which is why UV-B disinfection is used in some

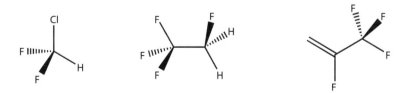

Figure 5.5 The chemical structure of common automotive CFC molecules. From left to right, we see R-12 (dichlorofluoromethane), R-134a (1,1,1,2-tetrafluoroethane), and HFO-1234 (2,3,3,3-tetrafluoropropene).

wastewater treatment facilities. We get polar ozone holes for a variety of reasons discussed in many other texts, but the dominant chemical cause is that chlorinated species tend to accumulate in a thin layer of liquid at the surface of stratospheric ice particles during the polar winter, and when the light returns and melts the ice in the spring, there is a spike in the concentration of chlorine radicals and rapid consumption of ozone.

Given the environmental problems associated with CFC refrigerants, many of the more destructive versions have been banned in industrialized nations. In their place, automotive air conditioning systems have shifted to the use of hydrofluorocarbons (HFCs), which are volatile methanes and ethanes with only fluorine and hydrogen attached to the carbon atoms. These structures maintain the unreactive nature and quality refrigeration properties of CFCs, but do not have the chlorine that causes damage to the ozone layer. Unfortunately, the CFCs and HFCs are also powerful greenhouse gases, which means that they easily absorb infrared radiation emitted by Earth at night, short-circuiting the natural planetary cooling mechanism and increasing atmospheric temperatures. For these reasons, additional refrigeration options are being considered, such as supercritical CO_2. However, the use of more-advanced refrigerant chemistries that are safe, prevent ozone destruction, and have a low greenhouse impact are still many years from implementation.

Refrigerant Controversy

Use of the refrigerant R-134a significantly reduced the role of automobile air conditioning systems in anthropogenic catalytic destruction of stratospheric ozone; however, this and many alternative refrigerants have enormous greenhouse gas potentials that impact climate change. It is known that most C–F bonds strongly absorb IR radiation between 1000 and ≈1400 cm^{-1}, and R-134a has an estimated 100-year global-warming potential of roughly 1200× that of CO_2. The various governmental environmental protection agencies and the automotive industry are looking for alternative refrigerant chemicals with lower greenhouse gas impact and recently approved a new hydrofluorocarbon refrigerant, HFO-1234yf, for use in passenger vehicles. HFO-1234yf is estimated to have several hundred times less global-warming potential than R-134a, though it still exceeds that of CO_2. It is also not compatible with existing R-134a refrigeration system designs, which means that auto manufacturers must completely redesign the refrigeration system before they can implement the new refrigerant. Based on the data supporting the reduced risk to the environment and preliminary safety testing, the European Union mandated the use of HFO-1234yf by European automakers by January of 2013 despite the design costs. However, Daimler (makers of Mercedes) and several other automakers recently backed out on this transition deadline based on new Daimler safety testing that shows HFO-1234yf may catch fire in head-on collisions, posing a potentially significant safety concern. The automakers are calling for a halt on implementation and the resumption of additional safety testing, setting up a battle between the auto industry and environmental groups. To the best of our knowledge, HFO-1234yf conversion remains unresolved at the writing of this text, and automakers continue to look at other alternatives such as CO_2 itself. Clearly, automotive-refrigerant chemistry and the delicate balance between performance, safety, and environmental stewardship are critical issues now and for the foreseeable future.

5.4 Braking and rotor types

Chemistry Concepts: thermodynamics, gas laws
Expected Learning Outcomes:
- Explain how brakes dissipate the energy of a vehicle
- Discuss the advantages and disadvantages of various rotor types from a heat-transfer perspective

Stopping a car involves removing the kinetic energy from the vehicle, and doing so quickly is essential from both a safety and performance perspective. Anything with drag or that generates friction is capable of robbing kinetic energy (or power) from a vehicle. Given enough time, the friction of the tires and in the engine, the drag in the transmission and differentials, etc., will fully dissipate the kinetic energy and bring the car to a stop. But a rapid stop requires a quick conversion of vehicular kinetic energy into some other form, followed by rapid dissipation of that energy in a safe manner. Modern disk brakes accomplish this conversion by turning the kinetic energy of the wheels into heat energy via friction, then giving this heat energy up to the atmosphere through molecular collisions using the thermodynamic principles discussed previously in this chapter. Combined disk and regenerative-braking technology is also starting to find applications in commercial vehicles. In regenerative braking, an electric motor used to turn the wheels is converted to a generator, and the resistance to turning an axle-mounted magnet within the generator coil converts some of the kinetic energy of the vehicle to an electrical energy that can be stored and used by the vehicle for another purpose. While regenerative braking is desirable from an overall efficiency perspective, this approach alone is not capable of stopping the vehicle as quickly as possible. At this stage, the best braking system for fast/emergency stopping is the disk-and-piston design that was mentioned when we discussed the gas laws in Chapter 1.

Understanding brakes involves physics more than chemistry and is thus outside the scope of this book, but we will briefly cover the sequence of events involved in stopping a disk-brake vehicle before discussing the role of chemistry in braking. Each wheel on your disk-brake vehicle has a metal disk called a *rotor* mounted between the wheel and the axel (Figure 5.6). In fact, the disk sits on the same studs that are used to mount the wheel, meaning that when the wheel turns, the disk rotates at the same angular velocity as the wheel itself. Surrounding the brake

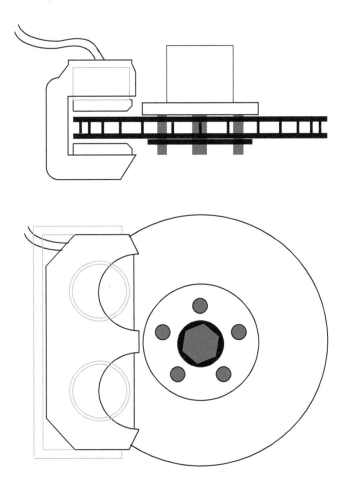

Figure 5.6 Diagram of the braking system in your vehicle from the top (top) and side (bottom).

disk is a C-shaped device called a brake caliper. The caliper and the parts mounted to it are fixed in space and do not rotate with the wheel. The caliper contains the piston(s) that actuate the brakes and has mounting points for the brake pads, which are small patches of friction material that make contact with both sides of the rotor when you press the brake pedal. Pressing on the brake pedal applies pressure to a hydraulic fluid that forces the piston in the caliper to squeeze on the brake pads, bringing them into physical contact with the rotor. When they make contact, the friction between the pads and the disk makes heat and wears away and/or vaporizes pad and disk material. A portion of the heat generated is radiated into the atmosphere from the materials themselves, and a portion is carried away by the layer of vaporized material. The harder you push on the pedal, the harder the caliper squeezes, and a greater amount of friction is generated at the pad-to-disk contact point. The higher the friction, the faster the car stops and the more rapid the heating, leading to a larger temperature change generated in the rotor and other brake parts during the stopping action.

One place where chemistry plays a crucial role in braking is in the friction material. Dry friction materials like those used in commercial vehicle brakes are composites that consist of friction inducers, binder, filler, lubricating agents, and possibly other components. Friction inducers are often strips of different metals or grains of hard materials like alumina, silica, or ceramics, though they can also be fibers of softer durable materials like aramids, Kevlar, carbon fiber, etc. Fillers are often inert and inexpensive materials like clay, which are a type of layered silicate material. Lubricating agents may be graphite or other layered materials that easily slide against one another due to weak intermolecular attractions between the layers (think of how smoothly your graphite pencil writes on paper). Other components of friction material may be chemicals designed to increase resiliency, like natural or synthetic latex rubbers (see Chapter 6). The binder is a place where chemistry is critical, as is the chemistry of the adhesive that holds the friction material to the brake pad.

Binders are often organic resins that easily volatilize during the braking process, modify the coefficient of friction and other mechanical properties of the composite, and maintain the structure of the material under the wide variety of temperatures and pressures experienced by the pad. These resins are often either phenolic in origin or resorcinols. Phenolic materials contain phenol, which is a benzene ring containing a hydroxyl group in place of one of the ring hydrogen atoms, while resorcinols are benzene rings that have two hydroxyl-for-hydrogen substitutions (Figure 5.7). Phenolic resins polymerize at relatively high temperatures via the reaction of phenol groups (C_6H_5OH) with formaldehyde (CH_2O), followed by dehydration reactions of hydroxymethylphenol ($HOC_6H_4CH_2OH$) with

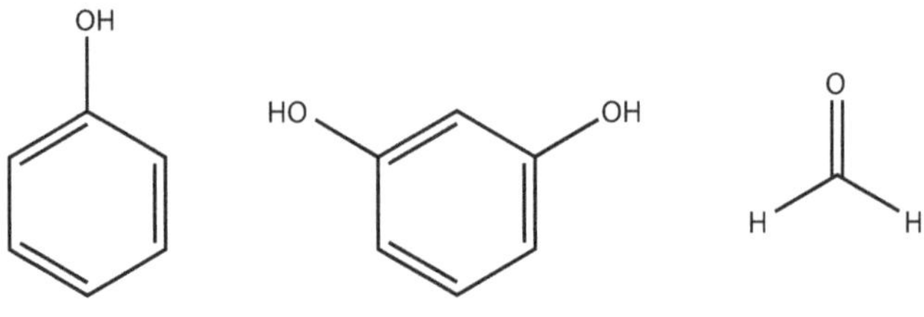

Figure 5.7 Structures of phenol (left), resorcinol (center), and formaldehyde (right).

other species to produce either methylene bridges (R–CH$_2$–R′) or ether bridges (R–O–R′) between phenols:

$$C_6H_5OH + CH_2O \rightarrow HOC_6H_4CH_2OH$$

$$HOC_6H_4CH_2OH + C_6H_5OH \rightarrow (HOC_6H_4)_2CH_2 + H_2O$$

$$2HOC_6H_4CH_2OH \rightarrow (HOC_6H_4CH_2)_2O + H_2O$$

The other components of the friction material can become either physically entangled in the polymerizing resin, or hydrogen bonds can form between the hydroxyl groups on the phenols and water or other H-bonding surface functional groups on the other components. Resorcinol chemistries are very similar, as resorcinol is a dihydroxy-substituted benzene.

The binder offers several potential chemistry-related sinks for the heat produced during braking, the dominant of which is the enthalpy of vaporization for vaporizing pad material. In phenol/resorcinol polymers, the relevant intermolecular attractions are likely to be hydrogen-bonding and dispersion-type interactions, which typically will require between 4 and 40 kJ/mol to overcome. This is well within the typical energy range of heat generated during braking. A second potential chemical sink for heat energy is in the breaking of chemical bonds in the binder. Typical phenol/resorcinol polymers contain C–C, C=C, C–O(H), C–O(C), O–H, and C–H bonds. The mean bond dissociation energies for these bonds are 317–376 kJ/mol, 728.3 kJ/mol, ≈460 kJ/mol, 208–264 kJ/mol, 361.9 kJ/mol, and 473.1 kJ/mol, respectively. It would take a very large amount of heat to induce breaking of these bond types, but potentially very high local heats may be generated near friction inducers in the friction material, making this a minor heat-dissipation mechanism. It is also possible that heating could lead to reactions between the organic material of the binder and oxygen in the atmosphere. This process is essentially combustion and requires only that we achieve high-enough temperatures to cross the activation-energy barrier for this type of reaction. Brake pad combustion typically is only a problem under very heavy braking for long durations, such as during long, steep descents of heavy vehicles in mountainous terrain.

Another place where chemistry is important in braking is in the rotor materials themselves. Much like the radiator materials discussed earlier in this chapter, the rotor must be capable of soaking up heat quickly and delivering it to the atmosphere quickly without cracking, warping, or suffering other mechanical failures. Sometimes rotors have specialized slots or holes drilled into them to facilitate more rapid cooling and dissipation of vapors generated at the point of contact with the friction material. Traditionally, both solid and drilled/slotted rotors are made of steel, but advanced-performance rotors and racing rotors are being constructed of ceramic or carbon fiber to save weight and increase high-temperature braking performance. Ceramics and carbon fibers are both capable of withstanding higher temperatures than steel without warping, but they exhibit very poor braking performance when they are cold. Thus, these advanced types of rotors are only found on high-performance cars and race cars. The design and chemistry of advanced rotor materials are discussed in many materials-engineering texts and research papers, and the interested reader is encouraged to consult such sources.

chapter six

Materials chemistry

Modern automobiles, both domestic and racing, have made numerous gains in performance, safety, and fuel economy in large part due to the use of advanced materials for automotive applications. Fiberglass and carbon fiber are very lightweight, yet can be made strong enough to replace both body panels and structural components such as suspension parts. They can also be designed to shatter in a serious accident, protecting the driver/passengers by dissipating the impact energy through all the fractures and flying debris. Likewise, these composite materials are not subject to corrosion the way that iron and steel are. Lightweight panels and low-rolling-resistance tires both contribute toward better fuel economy and overall vehicle efficiency. The design of advanced polymer membranes is helping to bring fuel cell operating temperatures into more suitable ranges for passenger cars, perhaps providing an alternative to fossil fuel–based combustion and grid-based electric vehicles. While the selection of materials may seem driven primarily by engineering concerns, there is a great deal of chemistry that goes into manufacturing these high-tech materials. In this chapter, we briefly discuss the chemistry behind many of the different types of lightweight or energy-conserving materials in modern automobiles.

6.1 Plastics and polymers

Chemistry Concepts: organic chemistry, intermolecular forces, polymerization reactions

Expected Learning Outcomes:

- Mention several advantages of plastic materials and where they appear in cars
- Explain stepwise and chain polymerization and characteristics of polymers produced by each method
- Define a copolymer and list at least three general types of copolymer

Plastic materials quite literally revolutionized the automobile industry, primarily because plastics offer many significant mechanical and manufacturing advantages over wood and metal components. They are easily moldable into complex shapes; they are almost entirely corrosion resistant; they can be produced in a variety of colors without the need for painting; they can be used in plating procedures to make lightweight chrome parts; and they resist denting and fracture. They can be made very rigid and strong like polycarbonates, soft and durable like polyurethanes, and either opaque or clear, depending on the degree of crystallinity in the polymer. In general, amorphous materials like glass are clear, since they lack long-range ordered arrangements of atoms that can block light. Some of the disadvantages of polymers include a lower resistance to temperature than metals and the possibility for photochemical reactions that may degrade either the polymer structure or pigmentation. However, the low cost, light weight, and workability of plastics often outweigh the disadvantages. Since plastics are organic polymers, polymer chemistry and definitions are required to discuss the chemistry of plastics and plastic production. Some basic polymer definitions and example polymers used in automotive plastics can be found in Appendix C. It is recommended that a reader unfamiliar with polymers read this section before continuing on in this chapter.

Polymerization is a chemical process used to connect polymer building blocks known as *monomers* through chemical bonds to build large chain-like molecules. There are two common general methods for generating plastics by this approach: *stepwise polymerization* and *chain polymerization*.

In stepwise polymerization, two monomer units can link at any time, chain growth can start anywhere at any time, and monomer disappears quickly. The average molecular weight, or the number of monomers per polymer, increases with time. The longer the reaction is allowed to proceed, the larger the polymer chains will get. Often, stepwise polymerization reactions occur via condensation reactions, which are reactions in which some atoms from each monomer are lost during the polymerization process. Many condensation reactions liberate water, which is obtained anytime a hydroxyl group reacts with an easily ionizable hydrogen atom, like those on carboxylic acid groups. Polyester, which is likely a major component of the fabric in your car seats, is a condensation polymer that forms ester (R–O–R′) linkages via an esterification reaction that liberates water. It is formed industrially by reacting terephthalic acid with ethylene glycol:

$$n\ C_6H_4(CO_2H)_2 + n\ HOCH_2CH_2OH \rightarrow [(CO)C_6H_4(CO_2CH_2CH_2O)]_n + 2n\ HOH$$

Since terephthalic acid is a diacid and ethylene glycol is a dialcohol, each polymer chain always has a carboxylic acid group at one terminus (end of the polymer chain) and an alcohol group at the other terminus, which allows the chain to grow at either end via the same water-liberating condensation mechanisms. This is an important component of stepwise condensation polymerization: The resulting polymer must have both important functional groups, one at each terminus, for the polymerization reaction to be effective. Other important polymers in your car formed via stepwise condensation reactions include the polycarbonate that is used in headlamp lenses and other parts and the polyurethane that is often used in seat foams and suspension bushings.

The second major type of polymerization mechanism, chain polymerization, generates polymers in a very different way. Polymers that result from chain polymerization grow via chain reactions that do not eliminate atoms from the monomers or polymers. Chain polymerization reactions form polymer chains rapidly, and the resulting polymers tend to develop similar chain lengths independent of the reaction time, in sharp contrast to the stepwise condensation reactions discussed previously. Chain polymerization reactions are sometimes called addition polymerizations, since the polymer chain grows without the loss of any atoms. Typically, chain polymerizations occur via a free-radical reaction mechanism and require the monomers involved to contain at least one C=C. Recall that a free radical is a molecule that has an unpaired electron. In chain polymerization, an initiator species is mixed with the monomers, and the reaction proceeds via a general multistep mechanism involving initiation, propagation/growth, and termination:

Initiation:

$$I \rightarrow R\cdot + R\cdot$$

$$R\cdot + M \rightarrow M_1\cdot$$

Propagation:

$$M_1\cdot + M \rightarrow M_2\cdot$$

Termination:

$$M_n\cdot + M_m\cdot \rightarrow M_n M_m$$

$$M_n\cdot + M_m\cdot \rightarrow M_n + M_m$$

$$M_n\cdot + M \rightarrow M_n + M\cdot$$

where generic monomer is represented as M, the initiator as I, and free radicals as species with dots behind the symbol. Initiation includes both formation of radicals from the initiator and reaction of the initiator radicals with monomer to make a monomer radical. Propagation is simply the reaction of active radicals on polymer chains with other monomers. Termination equates to the loss of the radical from the long polymer chain, and this can occur either by (a) the radicals at the ends of two polymer chains combining with or without addition of the two chains or (b) transferring the radical to a new monomer, which begins chain growth anew. Typical polymers formed by chain polymerization and used in automobiles include polyethylene, polypropylene, poly(vinyl chloride), and poly(methylmethacrylate), all of which are used to make clear and opaque panels in the interior of vehicle cabins, for example.

Another important facet of polymers is the idea of copolymers. A copolymer is a polymer chain that is made up of two different types of monomer/polymer, and the properties of the copolymer often differ from that of either pure component. There are several types of copolymer in which the order/size of the distinct fragments repeats in a regular fashion, including alternating copolymers where the monomer type changes each step in the polymer chain [(-A-B-A-B-A-B-)$_n$], periodic copolymers where there is a repeating sequence of monomer arrangements [(A-B-B-A-B-A-A-A-B)$_n$], and statistical copolymers where there is a certain probability of finding a particular monomer unit at the next point in the chain. There are also random copolymers where there is no repeating order to the monomer units along polymer chains and where the probability of finding a particular monomer at the next position is equal to the percentage of that monomer in the system. Yet another type of copolymer is the block copolymer, where pure polymeric chains are linked [(A-A-A-A-A-A-B-B-B-B-B-A-A-A-A-A-A)$_n$]. In the generic examples here, we consider only two types of monomer, but any number of monomers can participate in any one of the copolymer types. Another way of classifying copolymers is to consider whether the chains are linear or whether they are branched. Branching tends to improve adhesion via entanglement. Copolymers are often important in the rubbers used for tires and other components in your vehicle (see Section 6.3).

To make car parts from polymers, often the polymer has already been produced in a chemical plant and arrives as small, hard bits or beads of polymer resin made up of many physically entangled polymer chains. These bits and beads may also contain colorizing agents and other types of additives already, or the manufacturing plant can add these to the polymer. Most plastic car parts are made by melting the plastic beads, introducing the liquid into a form, and allowing the liquefied polymer to reharden in the desired shape. Plastics manufacturing is beyond the scope of this book, and the interested reader is referred to polymer texts and engineering books for a full description of possible manufacturing methods and additional details.

6.2   Composite materials

Chemistry Concepts: organic chemistry, solid-state chemistry, intermolecular (IM) forces

Expected Learning Outcomes:
- Define a composite material
- Describe the components of fiberglass and the origin of its properties

- Understand how carbon-fiber composite materials are produced and the origin of their properties
- Explain how using composite windshields improves safety
- Discuss the composite nature of clutch plates and other friction materials

Composites are materials made of two or more materials with very different physical properties that, when combined, produce a material with characteristics that differ from the individual component materials themselves. There are an enormous number of composite materials in an automobile. We have already discussed one example of composites when we covered the braking system: the brake pads. Some other examples of composites in your car include the fiberglass in Bondo and some lightweight body panels, carbon fibers for ultralightweight body panels and structural materials, laminated safety glass for car windows, and the clutch plates in manual and automatic transmissions. Rather than focusing on the mechanical pros and cons for these different composites and their applications, this section focuses on the chemistry used to make the materials.

Fiberglass is a composite containing glass fibers in a polymer matrix and typically involves a polymer that would otherwise be brittle, or one that fractures without deforming when subjected to stress. To fracture, a crack must form and propagate through the polymer at crystallite boundaries and other types of defects, which occurs readily in a brittle plastic. By embedding tough glass fibers in the plastic, a newly formed crack is likely to hit the glass fibers and be unable to propagate further through the composite structure. Effectively, the fibers help prevent catastrophic failure when the composite is subjected to moderate stress. At the same time, the strength of the glass fibers may lead to strength improvements over the unmodified polymer and impart some degree of flexibility to the otherwise brittle polymer. Yet under very high stress, fiberglass panels shatter into many small bits, which has the benefit of better dissipating energy in an impact, an important safety consideration for racing applications. As with most glasses, the glass fibers typically used in fiberglass are silica based (SiO_2), though a variety of natural inorganic materials are often included to reduce the working temperature (temperature at which the glass viscosity is low enough that it can be molded and manipulated easily) and impart chemical resistance to acid or alkali attack. The most common glass in fiberglass is known as E-glass and was patented by Owens-Corning in 1943. E-glass is an aluminoborosilicate glass formed by mixing the oxides of silica, boron, calcium, magnesium, and aluminum such that the melt is 52–56 weight% silica, 16%–19% calcium oxide, 3%–6% magnesium oxide, 12%–16% alumina, and 5%–12% boron oxide. This well-mixed melt is then forced through narrow nozzles and drawn into fibers. These fibers can be woven into glass fabrics or chopped into short segments of controlled length, with chopped fiber being more commonly used in automotive fiberglass. The polymer matrix in fiberglass is often composed of polyvinyl materials, polystyrenes, ester acrylate or methylmethacrylates, or acrylonitrile resins. Sometimes condensation-type polymers such as polycarbonates, polyesters, and polyphenylene oxides are also used as effective matrix materials. Whatever polymer is selected must develop strong adhesion to the glass fibers, meaning that it will (a) form strong intermolecular interactions with the oxide and hydroxide surface functional groups on the glass and (b) easily fill the spaces between fibers, thereby preventing the development of voids, or gas pockets, in the polymer matrix that would otherwise weaken the composite material. Typically, the matrix and fibers are combined together with solvents to make a viscous slurry. This slurry can be brushed or molded and heated to remove the solvent and/or cure the polymer, producing a solid composite panel.

Carbon-fiber composites are another class of composites with some commercial automotive applications and extensive use in race cars. Carbon-fiber composites also fall into the general category of fiber-reinforced polymeric materials. In general, carbon fiber offers many of the same advantages of glass fibers, but has some distinct advantages over glass-fiber reinforcement. For one, carbon has a lower molecular mass than silicon, meaning that carbon fibers will be lighter than glass fibers of the same size and shape. The surface of carbon fibers is also more likely to form strong interactions with hydrophobic organic polymers. At the same time, carbon fibers and glass fibers both develop very high tensile strength, leading to very strong composites with flexibility advantages over the pure-phase polymer. Most mass-production carbon fibers are formed by drawing either polyacrylonitrile or pitch melts in much the same industrial process as drawn glass fibers. In most automobile structures, the resulting carbon fibers are then woven into fabric sheets that are stacked and impregnated with the polymer matrix. Typically, the matrix is an epoxy resin, which is an organic resin that contains the epoxide functional group (see Appendix C). Common epoxy resins include phenolic-based resins and novolacs (Figure 6.1), both of which are heavily produced industrially. To polymerize and cure these resins, they must be mixed with a hardening agent that can open the epoxide ring and allow bonds between monomers to form, or they must be exposed to high heat. Organic compounds with reactive hydrogen atoms can be used for epoxy curatives, and most hardeners are amines or alcohols. By dissolving the epoxy resin in a suitable solvent, the viscosity can be reduced such that the woven fiber sheets can be dipped into the liquid and saturated with the resin. The wet sheets can be stacked and molded into forms, at which point heating finishes the curing process.

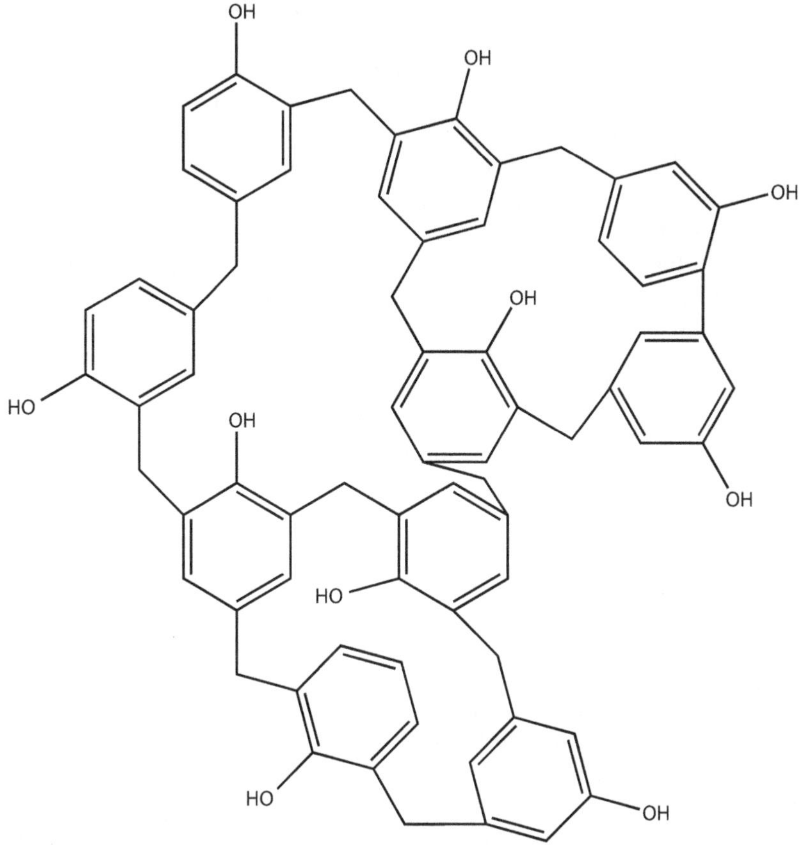

Figure 6.1 Novolac functional groups.

Monocoque Chassis

Early vehicles had welded steel frames that served as the key structural support, but over time, technology has moved to more advanced and safety-conscious structural designs thanks to advanced structural materials like those discussed in this section of the text. Perhaps the pinnacle of chassis strength and safety is the carbon-fiber monocoque chassis. A monocoque is a chassis where the external skin of the vehicle is part of the structural support for the vehicle itself. This eliminates the need to hang body panels on the chassis, which saves weight and also improves safety. In a collision, the outer panels of the car help to absorb the impact force and dissipate the energy more effectively than a frame-based or tubular chassis. As the structural panels crumple or shatter, they carry away energy that otherwise may injure the driver or passengers in a car more effectively than do panels hung on the frame at a small number of physical contact points, which only take up energy during fracture of the welds or other fixing devices. A steel monocoque is very heavy and would not be practical for racing applications; however, carbon-fiber composite offers similar strength at a much lighter weight. While monocoque chassis are very strong and safe, carbon-fiber monocoques are also extremely expensive. They are currently used in race cars, like the DW12 chassis used in Indy Car, and very high-end sports cars like the McLaren F1 and the Lamborghini Aventador. Shifting from unibody chassis in domestic vehicles to monocoques requires finding another structural material that balances the strength, weight, and price.

Automotive window glass is also a composite material known as a laminated composite. In automobile windows, a high impact resistance must be combined with rigidity and essentially 100% transparency (at least with respect to the windshield). Glass is an ideal material from nearly all perspectives, but glass tends to shatter and suffer brittle failure when its impact strength is exceeded. Likewise, polycarbonate plastics are also a great choice, but they too tend to shatter and suffer brittle failure when subjected to very high impact forces. The potential health hazard of shattered glass or polycarbonate traveling at high speed is a significant drawback to using these materials in car windows. However, if glass or polycarbonate is layered and bonded with a durable, flexible, and transparent polymer, then the polymer can help to absorb and dissipate the energy in the event of an impact and hold the glass/polycarbonate intact. This is precisely what is done when preparing laminated safety glass: layers of glass and clear polymeric sheets are stacked together and bonded. In the event of an impact, the polymer sheets prevent the glass/polycarbonate from shattering, while the glass/polycarbonate provides both strength and rigidity. Polymers typically used in laminated glass include polyvinyl butyric resin, aliphatic urethanes, or transparent cured resin sheets. These polymers make sense chemically because they contain oxygen and hydroxyl groups in their chemical structures, which should be able to form strong intermolecular attractions with the glass and polycarbonate functionalities, as well as imparting an ability to form chemical bonds via dehydration reactions during the bonding process. The polymers themselves are the adhesive that holds the various layers together, producing a layered composite material of glass and polymer.

Clutch plates in manual and automatic transmissions are much more complicated composite materials than the fiber-reinforced polymerics and laminated composites we just discussed, and are more analogous to the brake pads we discussed in Chapter 5. Clutch plates perform very similar roles to brake pads. They must generate a high coefficient of friction in certain situations, a very low coefficient of friction and low drag when not in use, and help to circulate a fluid to cool the component materials. In brake pads, the air and vaporized pad material serve as the cooling fluid, while in sealed transmissions the transmission fluid cools the points of contact between friction disks and other components. Most clutch plates have some type of grooving or channel pattern that can help circulate the fluid when not engaged and channel it out of the way when the friction disk is engaged. If a large amount of lubricant were trapped between the friction disks and gears or flywheel, the clutch plate would slip rather than allowing the gears to engage. Clutch-plate friction materials are composites formed from organic fibers, inorganic fillers and friction modifiers, polymers such as latex rubber, and an organic polymer resin that serves as the binder, much like brake pads. Most of the chemistry in the formation of clutch-plate friction material comes into play during the resin curing stage, when the solvent is evaporated and heat promotes polymerization of the resin and chemical interactions between the resin and friction material components. However, rubber containing friction materials also may involve galvanization chemistry, which is the subject of the following section.

A Tire for Every Season

Anyone who has purchased tires in a cold climate probably has noticed that there are summer tires, all-season tires, and winter tires on the market. A summer tire is designed for excellent performance when the weather is warm in both dry and wet conditions. Winter tires are designed to provide maximum traction when the weather is cold and the roadways are covered with ice or snow. All-season tires are designed with reasonable performance in mind over all the typical conditions of a year, but don't perform as well as summer tires in the summer or winter tires in the winter. But what precisely is the difference between these tire types that make them best during different times of the year? Many of the differences are engineering related. For example, winter tires tend to have deeper channels, squared edges to the tread, and siping (the small grooves in the tread blocks) that other tire types lack. All of these give snow tires more and sharper edge surfaces for biting into the ice and snow and to help move snow away from the contact patch. But there is also a big chemical difference that relates to the type of rubber and the ability to mix rubbers in the tire tread. There are several different elastomer building blocks used in tire rubber, including *cis*-1,4-polybutadiene, *cis*-1,4-polyisoprene, poly(isobutylene-co-isoprene), poly(styrene-co-butadiene), poly(ethylene-co-propylene), polychloroprene, and poly(butadiene-co-acrylonitrile). These different polymer and copolymer elastomers can be used in their pure form or blended together in the same tire structure to generate a mixture of rubber in the tread exhibiting targeted coefficient-of-friction and wear characteristics at a particular temperature. For example, the softness of the tread rubber is critical for traction and wear and is most easily related to the T_g, or the glass transition temperature, of the polymer. Polymers tend to behave much more like stiff crystalline materials below their T_g's. Typically, polymers do not exhibit rubberlike mechanical characteristics until they reach a temperature of $T_g + 30°C$ or hotter and they tend to get stiffer as the temperature increases. Thus, in a cold weather tire, one wants a rubber that includes elastomers with very low T_g's, such as *cis*-1,4-polybutadiene ($T_g = -100°C$) and natural rubber (*cis*-1,4-polyisoprene, $T_g = -72°C$), while warm weather tires benefit from inclusion of rubbers with higher T_g, such as nitrile rubber (polybutadiene-co-acrylonitrile, $T_g = -40°C$ to $-10°C$) and styrene-butadiene rubber (SBR; 50/50 polystyrene-co-butadiene, $T_g = -25°C$). Typical winter tires can generate a coefficient of friction on ice at $0°F$ that can range from 2 to 10+ times that of a summer tire at those temperatures. On the flip side, a winter tire composition in the heat of summer is likely to become stiff and wear away quickly. For more about rubber chemistry and tires, read on to Section 6.3 or read the 2006 National Highway Safety and Traffic Administration report,[*] which is freely available online.

6.3   Rubber

Chemistry Concepts: IM forces, organic chemistry, polymerization
Expected Learning Outcomes:
- List the common molecules found in natural and synthetic rubbers
- Describe the vulcanization process

[*] A. N. Gent and J. D. Walter, eds., "The Pneumatic Tire," National Highway Traffic Safety Administration, document DOT HS 810 561, 2006, http://www.nhtsa.gov/staticfiles/safercar/pdf/PneumaticTire_HS-810-561.pdf (accessed June 1, 2014).

- List common ingredients included in commercial automotive rubber products
- Describe the role of reinforcing fillers in tire technology

Cars use rubber in tires, hoses, on the surface of pedals, as gaskets to keep water out of the vehicle, in windshield wiper blades, and other applications. Some of these rubbers are very hard, like those that might be found in the suspension, while others are soft and easily bendable, like the gaskets in your door and the wiper blades. Some components have woven fibers or other structural materials sandwiched between layers of rubber, for example, the steel and fiber belts in your tires. The adhesion of the rubber to other materials as well as the physical properties of the rubber itself (melting point, elasticity, heat resistance, etc.) depend largely on the chemistry of rubber, which is the subject of this section.

Natural rubbers are latexes of polyisoprene produced by some trees and other plants. A latex is a stable suspension of polymer microparticles (particles with diameters of 100 nm to 100 μm), though in common English, *latex* has become synonymous with the general term *rubber*. Natural latex rubbers are often sticky, milklike liquids that range in color from pristine white to light brown. Since they come from biological organisms, natural latexes contain sugars, proteins, and other common plant biomolecules along with the polyisoprene in a solvent of water. Upon dehydration of the latex, the rubber microparticles coagulate and form a weak, soft solid with high elasticity and resistance to brittle fracture. Synthetic latexes are also produced industrially, and in these materials the organic polymer is either petroleum based or synthesized from natural feedstocks. Common synthetic latex polymers include styrene-butadiene rubber (SBR), acrylic polymers, and polyvinyl acetate, though the chemical rubber industry is now so advanced that many specialized rubber chemistries exist in the patent literature (see US Patent 6,613,838 B1 and the referenced patents therein).[*] A common feature of all the natural and synthetic rubber compounds is the presence of polyunsaturated carbon–carbon bonds (C=C) that are important for polymer stiffness and the types of chemical reactions in which the monomer and polymer molecules can participate.

From an automotive perspective, the most important chemistry involving rubbers is the vulcanization process. Vulcanization helps to make rubbers more rigid, more heat resistant, and imparts a great deal of strength by promoting the formation of chemical linkages between the polymer chains known as *cross-links*. The classic vulcanizing agent for natural rubbers is sulfur combined with high heat, though other chemical curatives with sulfur in the chemical structure (such as sulfenamides)

---

[*] J. Dove, "Synthetic Rubber Elastomers as Replacements for Natural Rubber Latex," US Patent 6,613,838 B1, filed Aug. 30, 2000, and issued Sept. 2, 2003.

can also be used for vulcanization. Pure sulfur vulcanization occurs very slowly, and as such, all of the vulcanizing agents can be combined with chemical accelerators to speed the vulcanization process to an industrially acceptable reaction rate. Accelerators include zinc oxide and stearic acid, though anything that can open the S_8 rings in sulfur and break apart the bonds in the sulfur chain will help accelerate the process. Effectively, these accelerators reduce the activation energy for cross-link formation in the rubber by providing a different mechanism for sulfur ring opening and bond breaking. Cross-links in the classic sulfur vulcanization process are actually chains of sulfur atoms that link together the organic polymers. While many theories exist about the detailed chemical mechanism of sulfur vulcanization, it seems clear that the presence of allylic hydrogen is critical (Figure 6.2). An allylic hydrogen is a hydrogen atom that is bonded to a carbon atom adjacent to a C=C. Removal of this allylic hydrogen, which is the easiest hydrogen to abstract in a nonconjugated alkene, forms a polymer radical that can then react with sulfur chains to form cross-links. The strength and heat resistance of the subsequent galvanized rubber is strongly influenced by the number of sulfur atoms involved in the cross-link. Short cross-link chains provide better heat resistance by having stronger bond energies, while long cross-link chains provide better resilience and lead to rubber with more flexibility by allowing more motional freedom for the polymer chains when under stress. Alternative curatives to sulfur that promote cross-linking of the polymer chains are also available, including peroxides, metal oxides, and urethanes.

Reinforcing fillers are also very important modifiers of automotive rubber that exert a strong influence over the performance characteristics. Traditionally, the most commonly used structurally reinforcing filler has been carbon black, which is also what gives the black color to vulcanized rubber. As an organic solid, the surface chemical functionality of carbon black leads to easy chemical interaction between the carbon particles and organic polymers in the rubber. Both C–C and C–S_n–C linkages between the filler and polymer provide the structural reinforcement, while the hardness of the carbon particulates provides abrasion and wear resistance, temperature management, etc. Thus, small, high-surface-area carbon blacks that can form many chemical linkages with the polymer provide the greatest benefits to vulcanized rubber. The other

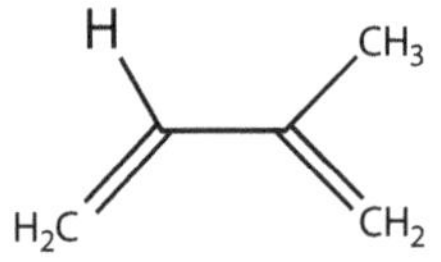

Figure 6.2 The lone hydrogen in the upper-left of the structure is an example of an allylic hydrogen.

most commonly used filler in the rubber industry is silica. Silica particles are harder than carbon black and can provide increased stiffness as well as lower rolling resistance, which improves fuel economy in silica-rich "green" tires. The surface functional groups on silica are mainly hydroxyl groups that do not interact easily with the polymer phase, since they more easily participate in hydrogen bonding or dipole–dipole and ion–dipole intermolecular (IM) forces, so the silica surfaces are often functionalized with organosilane compounds that are more chemically compatible with the polymer. Despite this surface functionalization, silica particles wearing at the rubber surface will likely re-form/reexpose hydroxyl groups, which can help improve the wet traction of a tire, as does the alteration of the tan δ, which is the energy dissipated during a stretch versus the energy released during relaxation to the original state. Both precipitated and fumed silica sources are used in the rubber industry; however, we note that functionalized silica is currently much more expensive than carbon black. This cost difference is primarily responsible for the fact that carbon black is the main reinforcing filler in automotive rubber applications today.

Green Tires

Tire commercials today stress fuel economy and "green" technology, but what is it that makes a tire "green"? There are two different strategies for making a tire more environmentally friendly. One is to improve the fuel economy of vehicles by reducing the rolling resistance of tires. This can be achieved, for example, by substituting silica or functionalized silica for carbon black as a filler, as discussed in this section. Michelin tire company estimates that 20% of fuel burned by a car goes to overcoming the rolling resistance of tires, and also suggests that rolling resistance in tires is responsible for up to 4% of anthropogenic CO_2 emissions annually. Many of the "green" tires advertising improved fuel economy for your vehicle are silica-modified rubbers. The other approach to green tires is to produce the rubber building blocks themselves from renewable and more readily available sources. Currently, rubber for tires comes from specific species of plants with a narrow global distribution or from nonrenewable fossil fuels. However, Goodyear and the biotech firm Genecor (now part of DuPont) have produced genetically engineered microbes that synthesize isoprene from sugars that are easily grown in a renewable fashion. The isoprene produced by the microbes can then be used to make synthetic polyisoprene, the same polymer extracted from rubber trees and produced synthetically from petroleum at Goodyear. BioIsoprene™ prototype tires have already been produced, and Goodyear expects to bring this type of "green" tire to market in the near future.

Another common class of synthetic rubber with automotive applications is the silicone-based rubbers and high-temperature sealants. Silicone rubber polymers are based on a backbone of silicon–oxygen bonds rather than the predominantly carbon–carbon or carbon–oxygen bonds that are found in latex rubbers and other organic polymers. The backbone in silicones is formed primarily via dehydration reactions of silanol groups that generate an –Si–O–Si– bond and liberate H_2O. Attached to this –Si–O–Si–O– backbone are a variety of organic functional groups, ranging from small methyl groups to aromatic phenyl groups. These organic functional groups have a great deal to do with the physical and chemical properties of the silicone rubber. For example, organic functional groups dangling off the backbone will help the otherwise hydrophilic siloxane chain be soluble in organic solvents and resistant to moisture. Most silicones are tacky, viscous liquids or gels that resemble glue in many respects before curing. Like the natural and organic rubbers, silicones must undergo a chemical process to harden or cure the rubber so that it holds its shape and can be handled more easily. Similar types of curing processes can be used in the organic and silicone rubbers, including vulcanization and the use of peroxide curing agents. Silicone rubbers are advantageous for many high-temperature applications, as they generally exhibit much greater thermal stability than their organic polymer cousins while offering many similar chemical properties thanks to the organic side chains. They also are generally available in a greater variety of colors, which appeals to the aesthetic taste of some automotive enthusiasts. Silicone rubbers are often used to make radiator hoses and gaskets in automobiles, bushings, and other components where organic rubbers are the industry standard. Silicone rubbers also offer good electrical resistance and are thus common materials for coating spark plug wires and other electrical components.

Over time, the rubber components of a car tend to break down as a result of physical wear or via a variety of chemical aging processes that lead to mechanical failure of the rubber. Most rubbers are subject to oxidation at C=C by ozone, which is a by-product of the tropospheric smog cycle. Eventually, this oxidation leads to a reduction in the average size of the polymer chain and can lead to cracking and a loss of integrity where tropospheric ozone is abundant. Likewise, oxidation of rubber by atmospheric molecular oxygen is also a problem. This type of chemical attack tends to form new oxygen-bearing cross-links between chains, leading to stiffening of the rubber. Additional stiffening with time occurs because the vulcanization chemicals and accelerators are still found in the rubber after curing. This means that the vulcanization reaction will continue at a slow rate over the life of a rubber part, leading to additional cross-links and stiffening of the rubber. Eventually, these mechanisms may make the rubber brittle and subject to additional cracking. Exposure to light of an appropriate energy can also alter the chemistry of tire rubber, both via the photolytic cleaving of chemical bonds (see Chapter 7) and by producing radical species in the rubber that can lead to additional cross-links or other undesired reactions. Since most of these chemical decay mechanisms involve oxidation, many tire manufacturers include antioxidant chemicals in their rubber formulations. Typical antioxidant chemicals in the rubber industry include amines, phenolic molecules, and organophosphite chemicals.

6.4   Alloys

Chemistry Concepts: materials chemistry, solid-state chemistry, crystalline solids

Expected Learning Outcomes:

- Explain what an alloy is and why it might be advantageous to use one
- Note several areas in a vehicle that benefit from alloys

Alloy and mag (magnesium) wheels are extremely popular among automotive enthusiasts and are frequently selected options on commercial automobiles. Certainly, custom or aftermarket wheels are an effective and relatively inexpensive way to personalize a car aesthetically, but what are alloy wheels, what do they have to do with chemistry, and what real advantages do they offer?

Alloys are solid solutions where the solvent is a metallic element. The solutes can also be metallic, as in the Pb/Sn alloy used in traditional electrical solder, or they can be a mix of metallic and nonmetallic elements, as in the Fe/C alloy we call steel. Binary alloys contain only two elements, while alloys in general may contain several different solutes. Most alloys are prepared by forming a melt containing the appropriate amount of the desired elements, and the composition of the melt is critical. If one looks at a simple binary solid–liquid phase diagram (Figure 6.3), you can see that at most compositions, one phase precipitates before the other, leading to a multiphase inhomogeneous system rather than a single homogeneous solid solution desirable in many applications. The special composition where the liquid mixture becomes a solid without intermediate phase precipitation is called a *eutectic composition*, and it is this composition that forms a high-quality solid solution needed for soldering and castable alloys.

On an atomic scale, the structure of the alloy is dominated by the crystal structure of the solvent metal. The solute atoms can be found in

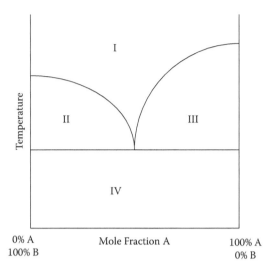

Figure 6.3 Generic binary liquid–solid phase diagram. In region I, both A and B exist in one liquid phase. In regions II and III, there is a liquid phase in contact with pure solid B and A, respectively. In region IV, both phases are in a solid. The only composition where the solid is homogeneous A and B is at the eutectic point, the point where all three lines intersect near the center of the image.

one of two places in the alloy crystal structure: either in the typically empty spaces between metal atoms called interstitial sites or as direct substitutions in which the solute appears in a position where we would ordinarily find a solvent metal atom. The location of the solute atoms depends primarily upon their atomic radius and the size of the interstitial spaces in the metal crystal lattice. For example, carbon atoms are small enough that they fit in the interstitial spaces between the iron atoms in the iron crystal lattice (C atomic radius = 70 pm versus 126 pm for Fe). On the other hand, in the Cu/Zn alloy that we call brass, the zinc atoms substitute for copper directly in the crystal lattice because they have similar atomic radii (128 pm for Cu vs. 133 pm for Zn).

The main advantage of forming alloys is that they exhibit chemical and/or mechanical properties that differ from the pure elements, typically in a positive manner. For example, adding carbon to iron results in a metallic substance we call steel, with much higher strength and toughness than iron alone. One might think that introduction of an element that tends to form covalent bonds such as C will change the nature of bonding in the alloy; however, iron and other alloys remain dominated by metallic bonding even in the presence of nonmetallic solutes. The performance advantage of steel versus iron arises because the interstitial carbon atoms reduce the free/unfilled volume in the solid and prevent movement of the crystals along defects that lead to plastic (irreversible) deformation. Likewise, adding chromium to steel improves the corrosion resistance of the alloy by allowing preferential oxidation of the chromium versus the iron. We call Fe–C–Cr alloys *stainless steel*. Steel and stainless steel are used for car body panels, structural components, exhaust plumbing, rotors, some types of wheels, etc.

The use of non-Fe alloys in automobile wheels strives from an effort to make cars lighter and reduce what is known as the *unsprung mass*. The unsprung mass is the weight of the vehicle that is not being supported by the suspension system, and reducing unsprung mass can improve vehicle handling, fuel economy, and the acceleration to some degree. Another benefit of Al, Mg, and related alloy wheels is that the metals have very high thermal conductivity, which helps to conduct friction-related heat away from the tires and the brakes. Making wheels from metals with a low molecular weight (such as aluminum or magnesium) accomplishes both of these tasks and was the original approach to making high-performance wheels. However, pure aluminum and magnesium are not as strong as steel and are subject to a variety of corrosion reactions that can damage the appearance or function of the wheels. The strength and corrosion resistance of aluminum or magnesium wheels can be improved by applying metallic or polymer coatings to the surface or by making wheels out of aluminum or magnesium alloys. Alloys are more desirable because if the surface coating is damaged on a pure Al or Mg wheel, water and ions will get under the coating and corrode the wheel from within (Figure 6.4). Magnesium-based alloy wheels typically contain 2%–12% aluminum along with lesser amounts of zinc, zirconium,

Figure 6.4 (See color insert.) Photo of aluminum wheel corrosion underneath a polymer wheel coating.

and/or more exotic metals. The mechanical improvement to the alloy arises primarily from the aluminum solute, while the other solutes provide corrosion resistance. Aluminum-based alloys often contain 2%–4% Mg along with Be, Mn, and Zn as antioxidants, and varying amounts of Si. The specific alloy formulations used tend to vary depending on the manufacturing method for producing the wheel (whether the wheels are cast, forged, milled, etc.). The special properties of alloys, particularly the corrosion resistance, make them abundant materials in automobiles.

6.5   Fuel cells: Power plant of the future?

Chemistry Concepts: quantum mechanics, thermodynamics, electrochemistry, organic chemistry

Expected Learning Outcomes:

- Explain the basic chemistry of an H_2/O_2 fuel cell
- Describe the chemistry of a Nafion membrane

As concerns grow about fossil fuels and the greenhouse effect, automakers rely on materials chemists and engineers to make important contributions to alternative vehicular power plants. Electric cars charged by efficient solar charging stations represent one highly green option, but this approach is still limited by the size and efficiency of current solar technology as well as our ability to store and recover energy in batteries and battery-like devices. Another alternative is to abandon combustion engines in favor of a power plant operating on a renewable fuel source that significantly reduces or eliminates CO_2 production, such as the hydrogen economy and electrical motors powered by automobile fuel cells. While numerous research and infrastructure challenges remain for making such a shift, several different research groups have recently developed materials that facilitate low-temperature splitting of water into hydrogen and oxygen, which indicates that a low-energy means of generating hydrogen may be feasible. Other materials groups have been working on developing safe hydrogen storage and transport mechanisms, such as reversible hydrogen storage in metal hydrides and ionic liquids (molten salts). Still other materials research groups are working on developing polymer fuel cell membranes that can function at temperatures appropriate for high efficiency in automotive applications. In this section, we discuss fuel cells, hydrogen production, and the current fuel cell chemistry that is most promising for fuel-cell-powered automobiles.

A fuel cell is an alternative to batteries for generating an electric current. In many ways, batteries and fuel cells are similar. In a battery, a chemical REDOX reaction occurs and generates a flow of electrons that we can use to do work. In a fuel cell, the same thing happens, but fuel cells have a continuous flow of fuel rather than the limited batch of fuel available in a battery. The ideal fuel cell for domestic automotive applications is the hydrogen/oxygen fuel cell, which reacts hydrogen gas and oxygen gas to make water and a current:

$$\text{Anode: } H_2 \rightarrow 2H^+ + 2e^-$$

$$\text{Cathode: } \tfrac{1}{2}O_2 + 2H^+ + 2e^- \rightarrow H_2O$$

$$\text{Overall: } \tfrac{1}{2}O_2 + H_2 \rightarrow H_2O$$

Each fuel cell of this type generates an electromotive force of 1.23 V.

The only by-product of this type of fuel cell is water, which would significantly reduce the automotive contribution to anthropogenic CO_2 production. Other fuels have also been explored, including methanol and other alcohols; however, these chemistries generate carbon dioxide as a by-product and are therefore less desirable:

$$CH_3OH + 3/2\ O_2 \rightarrow 2H_2O + CO_2$$

$$C_2H_5OH + 3O_2 \rightarrow 3H_2O + 2CO_2$$

All fuel cells contain an electrolyte, a fuel, catalysts at the anode and the cathode, and the electrodes/wiring. A wide variety of electrolytes exist for fuel cells, including liquids like the KOH fuel cells used in the space program, solid ceramics, and polymer electrolyte membranes. The polymer electrolyte membrane fuel cells (PEMFCs) are by far the most promising for vehicles, since they can be made into a very small and light package that operates at relatively low temperature. The polymer in PEMFCs must conduct protons to the cathode without conducting electrons. This is achieved by making the membrane a negatively charged ionomer, which is a neutral polymer backbone that has up to 15% negatively charged groups attached to the backbone as side chains. The most popular PEM to date is the Nafion membrane produced by DuPont. Nafion is a fluorinated copolymer with sulfonic acid groups. Nafion uses the sulfonic acid to facilitate proton transport via hopping of protons from acid group to acid group along the polymer matrix (Figure 6.5). The use of fluorine on the polymer backbone prevents any undesirable proton-exchange events between H^+ ions and the backbone. In principle, any ionomer membrane with sulfonic or phosphoric acid functionality would make a good PEM, assuming that the structure is stable at the operating temperature and that the proton exchange rates are fast enough to sustain the electrical current over the entire operating temperature range. In terms of the catalysts, most fuel cells require precious metals like platinum, nickel, and other transition metals to

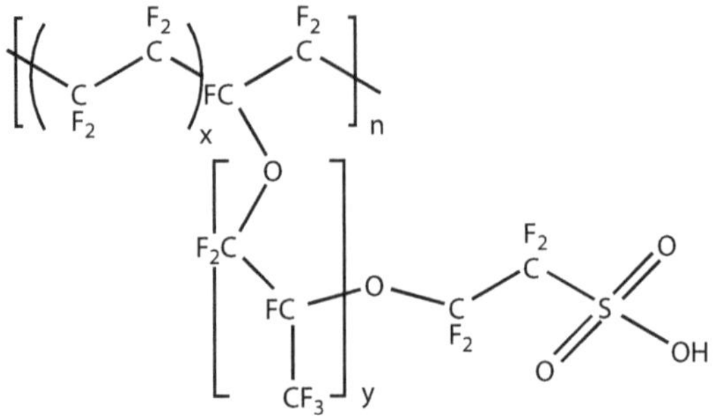

Figure 6.5 Structure of Nafion.

facilitate breakdown of the fuel at the anode and breakdown of oxygen/ recombination to water at the cathode.

To be useful in an automotive application, a fuel cell must be compact, produce a relatively high power output, and operate at low temperatures. Many of these requirements are at odds. For example, high power and high efficiency require operating the fuel cells as hot as possible, but fuel cells in cars must operate at temperatures down to −30°C (winter cold start) and may require some type of preheating. Humidity of the membrane is often critical, especially in the case of Nafion, which limits the operating temperature to one where the membrane will not dry out. Catalyst efficiency, poisoning, and costs are also a concern. In fact, there is not an efficient oxygen-splitting catalyst available to date, making fuel cell vehicles reliant on precious rare earth metals. Developing an efficient and targeted catalyst is an area of considerable research. Nonetheless, the 2012 U.S. Department of Energy goals for a vehicular fuel cell were an efficiency of 60% or greater at 25% net power output, drivable range of at least 250 miles, and a 2000-hour fuel cell durability, which were all achievable with prototype production vehicles at the writing of this book. The technology has come so far that Hyundai will be releasing fuel cell versions of the 2015 Hyundai Tucson in the coming year. However, issues remain regarding the cost of hydrogen production ($7–$13/gal gasoline equivalent), hydrogen storage and safety in the vehicles, the infrastructure cost to shift to hydrogen, and cost of the fuel cell and vehicles themselves.

chapter seven

Light and your car

Nearly every car lover has had a visual experience that has evoked strong emotion. Perhaps it was that moment when you saw your dream car painted flawlessly in your dream color at an auto show or when you witnessed that amazing burnout on the drag strip with six-foot flames bursting from the tailpipes. Our ability to experience these eye-catching sensations is driven by the ability of our eyes to convert colors of light, a type of radiant energy, into electrical signals that our brain can process and use to render an image. Sight is based directly on the interaction of light with color-sensitive chemicals in our retinas. Likewise, the interaction of light and matter is responsible for the myriad paint colors available and for the operation of high-intensity discharge lamps that make it safe to drive at night. Yet light–matter interactions are also responsible for breakdown of wax on your car, can potentially damage the paint, lead to yellowing and bleaching of plastics, and cause stiffening or cracking of rubber and plastic. The interaction of light and matter is also a fundamental component of modern chemistry, serving as the primary means chemists have to identify materials and their compositions, study molecular motion, and track the progress of chemical reactions. In this chapter, we introduce the basic concepts necessary to understand light–matter interactions and use them to understand the various ways that light plays a role in the properties of your vehicle.

7.1   Light and the atom

Chemistry Concepts: quantum mechanics, spectroscopy, atomic structure
Expected Learning Outcomes:
- Explain the concept of wave/particle duality
- Describe the electromagnetic spectrum
- Describe the fundamental process by which atoms and molecules absorb visible-light energy
- List the three possible outcomes of light absorption

Light is a form of radiant energy, which is energy that can be transmitted through empty space. Traditionally, light is considered to be a combination of perpendicular oscillating electric and magnetic fields; the so-called Maxwellian wave theory of light. Maxwell's electromagnetic light waves did an excellent job of explaining light behaviors like reflection, refraction, diffraction, and interference, and they are still used in many physics applications today. But near the start of the twentieth century, the wave theory of light had failed to explain the experimental observation of line spectra for the elements, the color and intensity of blackbody radiation (light emitted by a hot body), and the photoelectric effect (release of electrons from a metal surface when exposed to light). Max Planck, Albert Einstein, and Niels Bohr treated the interaction of light with matter, and later light itself, as occurring via small, quantized packets of energy called *photons* to explain, respectively, each of these phenomena. They showed that the energy of light photons is directly proportional to the frequency v through the famous equation $E = hv$, where the constant of proportionality h is called Planck's constant. Their work helped form the foundation of a new theory describing the atomic and subatomic world known as quantum mechanics. The modern view of light combines these two views, treating light as having both wavelike and particlelike properties. Because of wave–particle duality the Planck–Einstein formula can also be written in terms of the wavelength, which for waves is intimately related to frequency through $\lambda v = c$, where λ is the wavelength and c the speed of light in a vacuum. Therefore, as the wavelength of light gets smaller, the energy of that light increases, since wavelength is in the denominator when these two equations are combined: $E = hc/\lambda$.

Light is typically classified according to the energy of its photons in an organizational scheme called the *electromagnetic spectrum*. There are many processes in the universe that generate light with energies spanning more than 21 orders of magnitude when expressed in units of electron-volts (eV). Fortunately, many of the high-energy forms of radiant energy that strike the Earth are filtered out by the atmosphere or magnetosphere. Abundant forms of radiant energy at the Earth's surface include visible light, which is light with wavelengths of between 400 nm (violet) and 750 nm (red) that is detectable by our eyes, and some types of ultraviolet (UV) light, which is generally classified as light with wavelengths between about 100 and 400 nm. Visible light reflected from a car is what allows us to see a car and detect its color and finish. Ultraviolet light is responsible for many of the deleterious photochemistry phenomena (light-induced reactions) associated with car parts, as it has the energy required to break chemical bonds and form reactive radical species. We discuss both types of light in detail and how they affect your car in the subsequent sections of this chapter.

When light encounters matter, it can be refracted, reflected, or absorbed by that matter. Though reflection will be important when discussing color in the following section, chemists are most interested in what happens when matter absorbs light. Visible light absorption occurs primarily through electron excitation. For that reason, the novice chemist is strongly encouraged to read about atoms and atomic electron configurations in Appendix B, Section B.1 before continuing. Briefly, electrons live in relatively well-defined regions of space called *orbitals*. Electrons in outer, valence orbitals drive chemical and photochemical behavior. Electrons are allowed to move to other orbitals, and for an electron to jump from its lowest energy orbital to one of higher energy, it must acquire energy from outside the atom.

Electrons typically acquire the extra energy they need to transition to an excited state from light. When photons strike an electron with kinetic energy equal to or greater than the energy required to jump up to the next empty orbital, the electron can be bumped into an excited (higher energy) state. The photon that interacted with the electron is essentially gone, leading to a drop in the reflected/transmitted light intensity. Since each electron transition requires one photon and the probability of a photon–electron interaction can be calculated or determined experimentally, we can get some idea of how many atoms or molecules are in a chemical system by observing how much incoming light energy is lost as a function of wavelength when it passes through a sample. This is known as *absorption spectroscopy*. And since we establish in Appendix B that each atom or molecule has its own unique set of orbital energies related to the unique number and distribution of positive and negative charges, we can also identify substances or critical functional groups on a molecule by determining what light energies have been removed in our absorption spectrum.

It was suggested in Appendix B that an electron in an excited state will not remain excited forever. Since the most stable (or lowest energy) location for each electron is its original ground state position, thermodynamics says that the excited-state electron will eventually give up the energy it gained and return to the ground state. There are three common general methods by which a photochemically excited molecule gives up its excess energy. One possibility is a loss of excess energy as kinetic energy during collisions with other molecules, leading to a mean speed increase in a sample. This will increase the temperature of a system, so it is not unreasonable to say that, in this mechanism, the molecule gives up its excess energy to the thermal reservoir via a heat flow. A second possible outcome of photon absorption is that the molecule gives up the excess energy through a radiative transition. In other words, the molecule experiences a relatively sudden return to a lower energy state and emits a new photon of light that carries away the excess energy. The final possibility is that the excited-state molecule undergoes a chemical reaction. This can either be a rearrangement of atoms within the same molecule, splitting of a molecule, or reaction with a second molecule. If the atom gives up its energy via light emission, we can observe the emitted light via emission spectroscopy to count atoms or molecules, or to identify atoms or molecules based on the number and energy of photons emitted during the electron drop to the ground state.

7.2   Pigments and color

Chemistry Concepts: quantum mechanics, spectroscopy, atomic structure, thermodynamics

Expected Learning Outcomes:
- Explain the process that generates "color" when you look at a car
- Explain how one might chemically design an organic dye or a mixture of organic dyes to generate a particular color

To understand the chemistry that gives your paint color, we first have to understand what causes us to "see" the car in the first place. To see a car and note that the car is "red," photons of light must be interacting with the car in some way and reaching your eyes. We know that the car does not generate any photons on its own (assuming its headlamps are not on); otherwise we would always be able to see the car in full color when it is dark outside. This must mean that the photons detected by our eyes originate from our primary daytime photon source, the sun. If we remember that light can be refracted, reflected, or absorbed by matter, the fact that we see the car during daytime means that light from the sun must be getting reflected from the surface of the car and getting to our eyes. Light from the sun is called *white light* because it is made up of a continuous band of energies that fully span the visible-light range as well as higher and lower energy regions of the electromagnetic spectrum. A simple prism can show us this by spreading the white light out into its individual colors. Thus, the visible light from the sun making contact with the car contains photons of all colors, yet our eyes only see red photons when we look at the car, which implies that the car is only reflecting the red photons (Figure 7.1). What happened to the photons of other colors? The answer lies in the other light–matter interaction known as *absorption*, discussed in the previous section. Something at the surface of the car must be absorbing the other colors of light, and only the leftover red photons travel from the car surface to our eyes.

We call the substances at the car surface that absorb specific photons *pigments* and the critical light-absorbing functional groups *chromophores*. Pigments are chemical compounds that contain the appropriate electronic structure to absorb light of specific wavelengths, letting the surface reflect the remaining photons to our eyes. For example, if the car discussed here is to look red, the pigment molecules must be absorbing all the violet, blue, green, and yellow light photons. We can prove that this is related to the electronic structure of the pigment by delving deeper into the ideas presented in Section 7.1. There, we stated that visible-light absorption leads to excitation of electrons, but how do we know this? Niels Bohr showed this to be true in his analysis of the atomic line spectra in the early twentieth century. He was able to show that the violet, blue, and red lines in the emission spectrum of hydrogen result from electrons dropping from

Figure 7.1 (See color insert.) White light with all photon energies strikes the car, but only the red photons are reflected toward your eyes. The light waves in the image correspond to blue (shortest wavelength), green (intermediate wavelength), and red (longest wavelength).

higher energy orbitals to the second-highest energy orbital. His work used a special number n called a *quantum number* to indicate the ranked order of the orbitals based on their energy, with low n associated with low-energy orbitals and high n with high-energy orbitals. If we apply his key equation and the Planck–Einstein formula to explore the energy required for an electron to jump from the $n = 2$ to $n = 3$ orbital of a hydrogen atom, we find that the energy required is equivalent to that of 657-nm (red) light:

$$\Delta E = -R_H \left(\frac{Z^2}{n_{init}^2} - \frac{Z^2}{n_{final}^2} \right)$$

$$\Delta E = -2.178 \times 10^{-18} \ J \left(\frac{1^2}{3^2} - \frac{1^2}{2^2} \right) = 3.025 \times 10^{-19} \ J$$

$$\lambda = \frac{hc}{\Delta E} = \frac{(6.626 \times 10^{-34} \ J \cdot s) \ (3.00 \times 10^8 \ m/s)}{3.025 \times 10^{-19} \ J} = 6.57 \times 10^{-7} \ m = 657 \ nm$$

If you do many types of electron-transition energy calculations, you will find that very many of them occur in the 10^{-19}-J range, which puts the energy required within or close to the visible-light spectrum.

Since Bohr's predictions matched the experimentally observed wavelengths in the hydrogen emission spectrum, it confirmed that electron transitions between discrete orbital energies were responsible for the emission and absorption spectrum of hydrogen. Therefore, our red pigment must have portions of its electronic structure that tend to absorb light photons in the 400–600-nm range, leaving behind red.

There are a variety of organic chemical structures that give rise to color. For example, conjugated organic molecules have delocalized π electrons, or electrons that participate in the out-of-plane double bond, that can sorb photons of visible light. In fact, by adjusting the length of the conjugated carbon chain and/or by adding aromatic rings, the wavelength of visible light that most closely interacts with the pi electrons changes. In general, the longer the chain or the more aromatic groups in the structure, the longer the wavelength of light that the molecule will absorb. Organic pigments can also have their absorption properties tuned by including functional groups that have atoms other than carbon or hydrogen as appendages to the main molecular structure. Functional groups containing nitrogen (imine, azo, and nitro groups), oxygen (hydroxyl or carboxyl), sulfur, or replacement of hydrogen by halogens can all alter the absorption properties of the main structure. These groups are almost all more electronegative than C and H, meaning that their presence will pull electron density toward them. In essence, lengthening the conjugated carbon chain, adding aromatic rings, or replacing C or H with other atoms allows us to alter the shape and energy of the molecular orbitals. When the molecular orbitals arrive at an energy separation that generates a desirable color and that resists photochemical degradation (see Section 7.3), we have a pigment.

There are several families of both linear and polycyclic (more than one ring) compounds that span the full range of colors and are used in the automotive paint industry (Figure 7.2). Phthalocyanine pigments generate blue to yellow-green colors, depending on the extent of chlorine substitution for hydrogen in the base structure. Benzimidazolone pigments generate green to orange hues, depending on the types and extent of substitutions. Thioindigo pigments can range from purple to blue, while red and yellow colors are generated from azomethine complexes, flavanthrones, and isoindoline pigments. Both accumulated knowledge in industry and quantum mechanical predictions of electronic structure using computational chemistry programs can be used to determine the color a molecule will exhibit and/or to generate a particular color tone. A very detailed discussion of organic automotive pigments and their common synthetic pathways is available in the text by Herbst and Hunger.[*]

Another category of pigment is the inorganic pigments, or those that do not contain any carbon atoms but selectively absorb a fraction of

---

[*] W. Herbst and K. Hunger, *Industrial Organic Pigments* (New York: John Wiley and Sons, 2006).

Figure 7.2 (See color insert.) Typical categories of organic pigment used in the automotive industry. From left to right, starting in the top row: phthalocyanine blue, a benzimidazolone, thioindigo, a green azomethine, flavanthrone yellow, and a red–yellow isoindoline. Since many of these are chromophore families, the critical functional group for which the family is named is highlighted in red if the chromophore family is not variations on the entire structure. For flavanthrone, the anthrone group is highlighted.

visible-light photons. A majority of the matter on planet Earth is inorganic minerals, mostly oxides of metals across the periodic table. Most inorganic pigments involve crystalline oxides and salts of heavy transition metals that are low enough in the periodic table to have d-orbitals available, meaning metals in or below row 3. Typically, colorful inorganic oxides

Table 7.1 Inorganic Pigments and Their Associated Color

Color	Name	Compound
White	antimony white	Sb_2O_3
White	titanium white	TiO_2
Black	iron black	Fe_3O_4
Brown	raw umber	$Fe_2O_3 + MnO_2 + Al_2O_3$
Red	red ochre	Fe_2O_3
Red	cadmium red	$CdSe$
Orange	chrome orange	$PbCrO_4 + PbO$
Yellow	orpiment	As_2S_3
Yellow	cadmium yellow	CdS
Yellow	mosaic gold	SnS_2
Green	chrome green	Cr_2O_3
Green	viridian	$Cr_2O_3\,xH_2O$
Blue	Han blue	$BaCuSi_4O_{10}$
Blue	cobalt blue	$ZnAl_{2-x}Co_xO_4$
Purple	Han purple	$BaCuSi_2O_6$
Purple	manganese violet	$NH_4MnP_2O_7$

contain more than one type of metal. A short list of inorganic pigments and the colors they generate are provided in Table 7.1.

The types of electron transitions that typically absorb visible light in inorganic pigments include excitation to empty d-orbitals and charge-transfer reactions that involve changes in the metal oxidation state. In some types of coordination environments, the symmetry of the metal coordination shell generates an energetic splitting between the typically degenerate (equal energy) d-orbital electron energy levels. Often, these d-orbital splittings occur in the visible-light range. Colors are generated by electrons jumping between the two different energy d-orbitals. This type of mechanism is responsible for the blue color in cobalt blue ($ZnAl_{2-x}Co_xO_4$), where electrons jump between the two d-orbital levels of the cobalt. In charge-transfer reactions, visible light causes electrons to jump between atoms in a metal coordination shell and the metal themselves, changing the oxidation state of the metal. For example, $PbCrO_4$ experiences a charge-transfer reaction between the oxygen atoms and empty chromium d-orbitals that generates a yellow color. Inorganic pigments for automotive paints must consist of very fine particles so that the particles can be suspended in the paint solvent and generate a very thin, smooth coating on the panel surface.

Just like with any set of watercolors or model paints, mixing different types of pigments allows one to adjust the depth and tone of the paint color. Essentially, in a paint mixture, you are providing several different types of photon-sorbing mechanisms to the paint. Each unique pigment absorbs its specific types of photons and reflects others. Only the light wavelengths that are absorbed by none of the pigments are reflected to your eye at an appreciable intensity. For example, mixing yellow and blue pigments will generate a paint that has a green color. The yellow pigment removes the blue and violet photons from the incident white light and the blue pigment removes the yellows and reds. This leaves green as the only light photons that are reflected from the mixed pigment. The ratio of pigments will control the specific color of green that results by altering the efficiency with which the absorbed light photons are absorbed. Likewise, mixing in titanium white is a common method one can use to lighten a color, since you reduce the density of light-absorbing pigments and increase the density of light-reflecting pigments per unit surface area.

Finally, although they do not provide a chemical mechanism for changing the vehicle color, automotive paints often contain other components that affect what we observe. All of these ingredients either reflect or refract light in a manner that alters the appearance of the pigment. Metallic surfaces tend to reflect a large number of photons across the visible spectrum, giving the appearance of a very bright white-colored shine. Metallic-flake paint colors take advantage of the high reflectivity of metals by incorporating thin flakes of aluminum metal into the paint that give the dried paint a different reflectivity, depending on the orientation of the observer and light source. Metal-flake paints have a fine, grainy, sparkly appearance that is correlated with the size of the aluminum flake. There are also metallic paint colors that have a very high degree of shine, like a bare metal surface without the grainy effect of metallic-flake paints. These typically also contain aluminum flakes, but in this case the flakes are very homogeneous and very, very fine in size. This causes them to distribute evenly in the paint and along the vehicle surface, orienting such that your eyes do not detect the edges of individual flakes. Likewise, very small pieces of mica, a phyllosilicate mineral that generates atomically flat surfaces, and other layered materials can be added to paint to give an opalescent effect. Opalescence generally arises from an orientation-dependent refractive index change for the material (see following sidebar). Several types of paint modifiers provide this iridescent effect, including polymer microbeads and core-shell nanoparticles, but mica is used most frequently, since it is a cheap and naturally abundant material.

Color Flip Paints

Some of the most exciting automotive paints available seem to glow with color that shifts and changes with the viewing angle. Any paint that generates this effect falls into a category of special-effect paints called *interference paints*, which are thoroughly discussed in the book by Pfaff.* Essentially, these color-switching paints contain small, thin, and uniform flakes of an inorganic material coated with thin layers of one or more high-refractive-index materials. Since each layer and the substrate have their own refractive index, the angle of incident light penetrating the layers is altered at each interface. This changes the path length of light that passes through the layers and is reflected at the next interface with respect to light that is reflected from the particle surface. When the total path of the light passing through the different refractive-index coatings is equivalent to an integer number of the incident-light wavelength, the reflected and refracted light constructively interfere, allowing your eyes to detect photons of those colors. When the path length is not an integer number of the incident-light wavelength, then those light colors destructively interfere and are not observed when you look at the car. As your viewing angle changes, so do the angles of incident light that are reflected to your eye by each part of the car surface, imparting a spatial dependence to the paint color that changes as a function of your viewing position. There are many types of thin inorganic substrates used in color-flip-effect paints, including natural and synthetic micas, borosilicate glass, silica glasses, and others. The uniformity of the thickness and size of these particles combined with the ability to make the substrate flakes as thin as possible lead to more dramatic light-dependent effects. The high-refractive-index coatings are typically made of titania (TiO_2) or α-Fe_2O_3, though silicas and other inorganic oxides can also be used. Sometimes additional metal oxide films are included in the layering to help adhesion or to help direct the Ti or Fe oxides into one particular crystal structure, which gives the paint chemist additional control over the refractive index of the material. Controlling the refractive index and thickness of the various coatings allows one to control the dominant color/colors observed when one looks at the car. This is why there are several color-flip combinations on the market. In 2014, House of Kolors offers color-flip paints that oscillate between red/gold, cyan/purple, silver/green, green/purple, magenta/gold, gold/silver, and blue/red.

---

* G. Pfaff, *Special Effect Pigments: Technical Basics and Applications* (Berlin: Vincentz Network GmbH & Co KG, 2008).

7.3 Photochemical degradation

Chemistry Concepts: quantum mechanics, spectroscopy, atomic structure, thermodynamics

Expected Learning Outcomes:

- Define *quantum yield*
- Identify the bonds in waxes, pigments, and plastics most susceptible to photochemical attack
- Calculate the wavelength of light that can lead to photochemical degradation of a particular bond
- Describe the two mechanisms of photodegradation for organic polymers

Waxes, paints, coatings, and interior/exterior plastics are composed wholly or primarily of organic molecules and are subject to UV-based attack because UV light has energies similar to those of common bond types in organic polymers. (See Table 7.2 and Appendix C for a list of common connectivities in organic compounds.) This means that UV light can break bonds in automotive polymers, causing permanent chemical changes as opposed to the harmless electron transitions that accompany the sorption of visible light. Both photochemical bond breaking and the subsequent chemical reactions that can happen after a bond is broken change the molecular structure of a polymer, leading

Table 7.2 Mean Bond Energies for Typical Bond Types in Automotive Organic Molecules and the Wavelength of Light Corresponding to That Bond Energy

Bond type	Mean bond energy (kJ/mol)	Cleavage λ (nm)
C–H	413	290, UV-B
C–C	348	343, UV-A
C=C	614	195, UV-C
C–O	358	334, UV-A
C=O	799	150, UV-C
C–N	293	408, violet/UV-C
N–N	163	734, red/IR
N=N	418	286, UV-B
N–H	391	306, UV-B
C–S	259	462, blue
S–S	266	450, blue

Source: T. L. Brown, H. E. LeMay, B. E. Bursten, C. J. Murphy, P. M. Woodward, and M. W. Stoltzfus, *Chemistry: The Central Science*, 13th ed. (Upper Saddle River, NJ: Pearson Education, 2014).

to loss of color, a change in physical properties, and/or mechanical failure. These deleterious chemical processes initiated by photon absorption are known collectively as photochemical degradation. In this section, we examine the details of UV light–matter interactions that lead to photochemical degradation of organic polymers in and on your car.

Before discussing details of photochemical degradation, we must first revisit the concept of light absorption. As noted earlier, an atom or molecule that is rotationally, vibrationally, or electronically excited can give up its energy as heat through molecular collision, as light via an emissive radiative transition, or by using that energy to participate in a chemical reaction. It is important to realize that although each of these outcomes is possible when a photon absorption occurs, not all of the outcomes occur with the same likelihood. To capture the efficiency with which a particular energy release occurs, chemists use a term called *quantum yield*. The quantum yield is the number of times a particular outcome occurs divided by the total number of photons that are absorbed. For example, if every photon absorbed leads to a photon emission of energy X, the emission X quantum yield will be 1 and both the heat and reaction quantum yields will be zero. Photodegradation falls within the chemical reaction category, since it involves the breaking of chemical bonds by light. Thus, for photodegradation of an automotive polymer to be a significant concern, (a) the polymer must be exposed to light whose energy exceeds the energy of a bond in the structure and (b) bond breaking must have a high quantum yield.

To prove that sorption of UV light can cause photochemical degradation of automotive polymers, let us consider the energy of typical bonds found in waxes, paints, etc. Organic molecules in automotive polymers are dominated by C–C, C=C, and C–H bonds, though it is common to find –C–O–C–, C–OH, C=O, and carboxylic acid/carboxylate bonds as well. Other bond types found in various classes of pigment, plastic, and automotive rubbers include C–S, S–O, S=O, C–N, S–S, N–H, N–N, and N=N bonds. We can identify the wavelengths of light that can break each type of bond by considering their mean (average) bond energies. For example, if we examine a typical C–H bond using the data presented in Table 7.2, we can apply the Planck–Einstein formula written in terms of wavelength rather than frequency to determine the light wavelength required to break a typical C–H bond:

$$E = \frac{hc}{\lambda}$$

$$\therefore \lambda = \frac{hc}{E}$$

$$\lambda_{C-H} = \frac{hc}{E_{C-H}} = \frac{(6.626 \times 10^{-34} \text{ J} \cdot \text{s}) \times (3.00 \times 10^{8} \text{ m/s})}{413,000 \text{ J/mol} \times \dfrac{1 \text{ mol}}{6.022 \times 10^{23}}} = 2.90 \times 10^{-7} \text{ m} = 290 \text{ nm}$$

Avogadro's number (6.022×10^{23}) is used in the denominator because we require the energy of an individual bond rather than the molar bond energy for this calculation. The results of similar calculations for the other bond types are also presented in Table 7.2. Clearly, even when one considers the energy variations likely for each bond type, depending on the specific details of the molecular structure, light in the UV-A and UV-B regions of the electromagnetic spectrum are high enough in energy to be a significant problem for the organic polymers in or on your car. UV-A and, to a much lesser extent, UV-B are also regions of the electromagnetic spectrum that are not terribly well filtered by our atmosphere, meaning that relatively high quantities of these photons can be found at the Earth's surface.

UV-A and UV-B photons can initiate two main types of photochemical processes that lead to degradation of organic polymers in cars, namely photolysis and photooxidation. Photolysis refers to the breaking of a chemical bond due to sorption of a light photon:

$$R–R' + hv \rightarrow R\cdot + R'\cdot$$

Photolysis therefore breaks polymers apart into smaller molecular fragments that each contain an unpaired electron, altering the physical properties of the bulk material and the electronic structure of the molecule in question. These molecular fragments bearing the unpaired electrons are known as *free radicals,* and as has been stated several times previously in this book, free radicals are very reactive species in chemistry. These free radicals can initiate other non-photon-based chemical reactions in the polymer that alter its properties. For example as we learned in Chapter 6, free radicals are quite capable of attacking C=C bonds, leading to loss of the double bond and addition of the radical species at one of the associated carbon atoms. When this occurs, the molecular weight increases and the molecules in the polymer become more highly branched. If a polymer forms via a chain-polymerization mechanism, these new radicals may also lead to additional polymerization in the plastic, rubber, or coating, eventually leading to stiffening or cracking as the materials become more rigid and crystalline. Whether photolysis leads to a reduction or increase in the mean molecular weight of the polymer, photolysis and the subsequent chemistry will alter the electronic structure of the coating, affecting its absorption spectrum and leading to photobleaching of the pigment or plastic.

The second major type of photon-initiated degradation reaction is photooxidation. In this mechanism, the photon sorption by the organic polymer initiates a series of reactions with atmospheric oxygen that lead to oxidation of carbon atoms to form peroxy radicals (ROO·) and hydrogen peroxy molecules (ROOH):

$$R–R' + h\nu \rightarrow R· + R'·$$

$$R· + O_2 \rightarrow ROO·$$

$$ROO· + R'H \rightarrow ROOH + R'·$$

These new peroxy species can undergo further photon-induced reactions to produce aldehydes and ketones due to the photochemical instability of the peroxide bond. Likewise, oxy radicals and hydroxide radicals are produced:

$$ROO· + h\nu \rightarrow \text{aldehyde} + CH_2 = CH–R' + ·OH$$

$$ROOH + h\nu \rightarrow RO· + ·OH$$

$$ROOH + h\nu \rightarrow \text{ketone} + CH_2 = CH–R' + H_2O$$

Just like in the chain-polymerization mechanism discussed in Chapter 6, any of these radicals can terminate by a mutual termination mechanism, leading to ether linkages (R–O–R') and other new chain chemistries:

$$RO· + R'· \rightarrow R–O–R'$$

A much more detailed discussion of organic polymer photodegradation is provided in the text by Rabek,[*] and those with more advanced chemistry experience are referred to this source. Just as for photolysis, the net result of photooxidation will be a polymer matrix containing new molecular structures with different orbitals than the parent material. These changes will impact their ability to sorb photons and will alter the color of or completely bleach away pigments. What is different about this particular photodegradation mechanism is that the increase in oxygen functional groups will make the polymer more hydrophilic, a significant problem for waxes and other coatings designed to protect parts from interacting with water.

To help prevent both of these photo-initiated degradation mechanisms, many polymers and rubbers contain protective chemicals that filter UV light, capture radicals, and/or break down peroxides.

---

[*] J. F. Rabek, *Polymer Photodegradation: Mechanisms and Experimental Methods*. (New York: Springer, 1995).

The purpose of UV absorbers is to capture the high-energy UV photons before they can interact with the bonds in the polymer. There are several efficient organic and inorganic UV absorbers available. Metal oxides of lanthanide or noble metals having good UV-absorbing properties are frequent inorganic UV absorbers, while organic UV absorbers include aromatic molecules with relatively high contents of nitrogen and oxygen, such as oxyanilines. Radical-capturing organic molecules include sterically hindered phenols and amines that can easily give up a hydrogen atom and rearrange to form stable nonradical molecules. Organic molecules with sulfur- or phosphorus-bearing groups where the sulfur and phosphorus are capable of being oxidized are good peroxide destabilizers. Each of these protectants can be included in a paint, rubber, or coating in a limited quantity, but when they are used up, photochemical degradation can proceed more rapidly.

Another type of polymer degradation that is indirectly related to light is the thermal degradation of wax coatings and interior plastics. As noted earlier, one mechanism for eliminating the extra energy when a molecule absorbs light is to dissipate that energy as heat, which effectively increases the temperature of and molecular motion in the surrounding molecules. Many dark-colored paints and plastics absorb a significant number of visible-light photons per unit surface area, generating very high surface temperatures. For example, placing your hand on a black car's body or on a black rubber steering wheel after the car has sat in the summer sun for an hour usually results in burns or an inability to hold the wheel. Surface temperatures on body panels and in polymers can be extremely high as a result of this energy dissipation via heat—up to 195°F. Unfortunately, this sort of temperature is well within the melting-point range of many natural waxes like beeswax and carnauba wax as well as many synthetic carnauba-containing blended waxes. The subsequent liquefaction can cause the wax to run off the car body, vaporize, or be more easily removed by water from rainfall or washing a car. This is the reason you need to apply carnauba waxes more often and why many car-care products tell you to apply waxes or to wash cars in the shade. Likewise, the hot interior temperatures of vehicles affect plastic parts by volatilizing remaining monomers in interior plastics, giving cars the "new car smell" and the "very used car smell" that you have assuredly experienced in the past. The heat can also lead to softening of the plastic parts, making it easier to deform or scratch them. Though none of these processes are instances of direct photochemical degradation of bonds, like the UV degradation discussed at the start of this section, thermal degradation of waxes and plastics is an indirect result of visible-light photon absorption by organic molecules.

7.4   Headlamps and LEDs

Chemistry Concepts: quantum mechanics, thermodynamics,
   semiconductors
Expected Learning Outcomes:
* Explain the operation of a common halogen headlamp
* Describe the halogen cycle and how it prolongs bulb life
* Describe the difference between a common halogen lamp and a
  high intensity discharge (HID) lamp
* Describe how diodes work and how they generate light

Automotive headlamps and taillamps produce light via a number of
mechanisms strongly rooted in chemistry. Recall that light can be generated either via luminescence of a hot body, known as *blackbody radiation*, or
by stimulating electron transitions in chemical systems. Blackbody radiation often requires a controlled chemical environment to avoid oxidation
of the body (the filament in the case of a lamp) and oxidative fracture at
the high temperatures required for luminescence. Emission-type lamps
rely on the identity of the chemical system and the available electron transitions to produce light of a specific color, and they must contain a mechanism that can stimulate those transitions. Both types of light-producing
mechanisms have been used in automotive headlamps, and headlamp/
taillamp technology is beginning to use semiconductor light-emitting
diodes (LEDs) as light sources. In this section, we step through the chemistries of the light sources in modern automobiles, including halogen
lamps, high-intensity discharge lamps, and light-emitting diodes.

Halogen headlamps are quite similar to the incandescent lamps that
you may have in your home, but they also incorporate an ingenious
chemical method that extends the life of the filament. In conventional and
halogen headlamps, there is a wire filament typically made of tungsten
connected to two electrodes. The bare wire is a source of resistance, and
when one forces a flow of electrons through the wire, the filament heats
and, when hot enough, begins to emit radiation. The tungsten filaments
in a halogen headlamp typically operate at temperatures above 2500 K to
produce white light via a blackbody-type luminescence mechanism. The
major problem with the resistance-filament style of headlamp is chemical: at this high operating temperature, the filament easily oxidizes in the
atmosphere and breaks. Conventional, halogen, and incandescent lamps
combat the oxidation process by backfilling the bulb with an inert gas
(either nitrogen or a noble gas), thereby excluding oxygen and extending
the lifetime of the filament. However, there is another drawback to fil-
ament-style lamps: at their operating temperatures, they vaporize some
of the filament atoms, which eventually reach the bulb surface and condense. The condensed metal atoms make a black coating on the bulb

interior that eventually covers enough surface area to block the light, even if the filament lifetime has not been exceeded. Here is where the halogen headlamp has an advantage. By introducing a small amount of bromine or iodine gas to the bulb along with the inert gas, a chemical reaction occurs that clears the deposited tungsten coating and returns the evaporated tungsten atoms to the filament. The hot halogen gas travels through the bulb to the wall where it reacts to form tungsten iodide or tungsten bromide:

$$W + 3I_2 \text{ (g)} \rightarrow WI_6$$

At the operating temperatures of the bulb, these compounds are in the gas state and easily diffuse or move around within the bulb. When the halogen salt vapor reaches the filament, the temperature is hot enough to dissociate the tungsten–halogen bonds, regenerating the iodine or bromine gas and redepositing the tungsten on the filament.

The brightness of an incandescent light is related to the operating temperature: the hotter the filament, the brighter the light. However, higher operating temperatures also lead to faster evaporation rates of the filament and decreased lifetimes. The presence of the halogen extends the bulb lifetime and permits the bulb to operate at higher temperatures because it regenerates the filament rapidly to combat the fast evaporation rates while also removing the opaque surface coating. For these reasons, the tungsten halogen lamp remains the industry standard for automotive headlamps.

Many manufacturers, especially high-end luxury carmakers, have opted in recent years for a brighter bulb technology known as a high-intensity discharge (HID) lamp. You can recognize HID lamps by their bluish tint. HID lamps operate on a completely different principle than incandescent halogen lamps. HID bulbs are more similar to the discharge tubes you may have observed in general chemistry (when discussing line spectra) than they are to halogen bulbs. A HID bulb consists of a noble gas (typically either Xe or Ar), a mixture of halide salts, and what is called a *burner*. The burner is simply two electrodes separated by a gap. When the lamp is on, high current is passed through the electrodes such that an electrical arc is generated. The arc in turn heats the noble gases and salts in the bulb, vaporizing the salt and causing the noble gases to form a plasma—a charged gas-phase atom or molecule. The plasma helps conduct electrons between the electrodes, stabilizing the electrical arc in the burner. Though this stable arc produces some light, the primary role of the arc is to stimulate electron transitions in the plasma and the vaporized salts. Relaxation of the excited-state electrons in the salt vapors and plasma molecules back to their ground states leads to emission of light, and the visible light generated in these events combines to produce the vast majority of the light produced by a HID bulb. HID lamps operate at higher temperatures than incandescent halogen lamps (plasma temperatures of 5000–5500 K, which is easily high enough to vaporize the halogen salts) but require a brief start-up period, since the salts are in a solid phase until their vaporization temperature is reached.

Argon bulbs have a longer warm-up cycle and will spark, dim, and then slowly build in intensity when started. Xenon bulbs tend to reach maximum luminosity almost immediately after start-up. This is related to the relative first ionization energies of these noble gases: Xe has a much lower ionization energy than argon (1170.4 kJ/mol versus 1520.6 kJ/mol) and thus forms a stable high-temperature plasma more quickly. The salts are nearly always iodides, though the anion can be any relatively volatile halide, and most HID bulbs contain a mixture of indium, thallium, and sodium to produce a broad light spectrum. Indium (In) emits light in a distribution around 410 nm and 451 nm (purple/blue); Th emits light in a distribution around 535 nm (green); and Na emits light in a distribution around 589 nm (orange/yellow). Some lamps also contain zinc, yttrium, scandium, and other exotic metal halides to optimize the spectral output or operation/stability of the burner. The bulbs themselves or the lenses may also contain coatings or inclusions of organic dyes or inorganic photoluminescent materials, such as transition-metal coordination compounds, that alter the spectrum of the discharge lamp to achieve certain criteria (color temperature, spectrum, etc.). A fairly detailed list of potential spectrum-modifying compounds for HID lenses is available in US Patent US20040095779.[*] HID lamps offer better brightness at a lower power consumption than halogen bulbs, but they are significantly more expensive and tend to produce more glare that can distract other drivers.

Light-emitting diodes (LEDs) are starting to become common in taillamps and are currently being optimized for use as headlamps. The use of these devices as automotive light sources was pioneered in the Audi R8 Le Mans race car. LEDs can be designed to produce very bright light at very low power like the HID lamps, but they are comparatively inexpensive. Like HIDs, LEDs produce light based on electron transitions, but in this case, the wavelength of light emitted is related to the energy difference between the valence and conduction bands of a semiconductor pair, which is called the *band gap*. A semiconductor is a material that can serve as an insulator or a conductor, depending on the conditions. Often, semiconductors useful for light generation or capture are formed by bringing two chemically doped materials into contact such that one material is deficient

---

[*] D. Bryce, P. Schottland, and B. Terburg. 2004. "Automotive Headlamps with Improved Beam Chromaticity," US Patent 20,040,095,779 A1, filed Oct. 3, 2003, and issued May 20, 2004.

in electrons and the other rich in electrons. In chemistry, a deficiency in electrons generates a positive charge, and perhaps not surprisingly, we can generate the electron-deficient part of the semiconductor pair by including a higher valence impurity in the parent material. For example, allowing P^{5+} to substitute for Si^{4+} in a silicon-based semiconductor gives an excess of positive charge and forms what we call a *p*-type semiconductor. We often say the excess positive charge generates "holes" in this material. The other member of the semiconductor pair is called an *n*-type semiconductor and contains an impurity of lower valence that leads to an excess of electrons with respect to the base material, for example, B^{3+} substituting for Si^{4+} in a silicon-based semiconductor. Application of a current to this pair when they are in physical contact causes electrons and holes to flow toward the point of contact, termed the *p/n* junction, where they can combine with one another. In an LED, the electron/hole recombination at this junction generates light. The chemical composition of the semiconductor and chemistry of the dopants is what determines the size of the band gap in these materials and, thus, the color of the diode. This means that chemistry can be used to produce LEDs that emit light at nearly any visible wavelength.

The previous discussion addresses the light emission or "LE" part of LED, but what exactly does the "D" mean? In electronics, materials that permit a flow of current in only one direction are called diodes. Ideally, think of a diode as zero resistance one way and infinite resistance in the other, much like driving on a one-way road. The nature of the doped semiconductor pair allows current to flow from *p*-doped material to *n*-doped material, but not in the other direction, making the pair of light-emitting semiconducting materials inherently an electrical diode.

The final critical feature that makes an LED different from a general semiconducting material is that the electron-hole recombinations must result in radiative transitions, or transitions that emit light. Simple silicon and germanium semiconductors like those often used in the computer industry have nonradiative recombinations. More complex semiconductors that are various combinations of gallium, arsenic, aluminum, indium, phosphorus, and nitrogen are used for visible-light diodes. The two main obstacles for using LED lighting in automotive headlamps and taillamps seem to be (a) removing the heat generated by the LED and (b) overcoming inherent optical issues with LEDs. Both of these challenges have more to do with physics than chemistry, and as such these are not discussed here.

Appendix A: Matter and measurement

A.1 Properties and classification of matter

Chemistry Concepts: matter, chemical and physical properties, chemical and physical changes

Expected Learning Outcomes:

- Define *matter*
- Define *element* and *compound*
- Distinguish between physical and chemical properties
- Distinguish between physical and chemical changes

Matter is anything that has a measurable mass and volume. Matter is everywhere (at least here on Earth). Even the air in your car's tires contains matter. Matter (such as air) cannot necessarily be seen with the naked eye, but in every case its mass and volume can be determined either via direct or indirect measurement.

At a smaller scale, matter can be considered to be anything that is made of atoms or molecules. Atoms and molecules are the smallest units of a substance that retain its chemical behavior. To understand this further, we need to define these terms and a few others. In a general sense, we consider atoms to be basic building blocks of matter, even though atoms contain even smaller (subatomic) particles, in particular protons, neutrons, and electrons. The identity of an atom is determined by the number of protons it has. Protons have a positive charge and a very small mass and are found in the nucleus, or center, of an atom. Neutrons are also in the nucleus and have a similar mass to protons, but have no charge. Together, protons and neutrons make up the bulk of the mass of an atom, but occupy a very, very small volume ($\approx 1/10{,}000$ the volume or less). Surrounding the nucleus are electrons, which have a negative charge and far smaller mass than protons or neutrons. The number and arrangement of electrons in an atom determine many characteristics, including its size (radius) and reactivity.

An element is a pure substance that contains one type of atom. For example, one atom of carbon is the smallest unit of the element carbon, in which each atom has exactly six protons. They may vary in the number of neutrons, which affect their mass. When atoms of the same element vary in the number of neutrons, they are called *isotopes*.

Atoms react with one another and combine to form molecules. A molecule is the smallest unit of a compound that retains its chemical behavior. For example, one molecule of ethanol is the smallest unit of the compound ethanol, and all ethanol molecules have the same chemical formula, C_2H_5OH. Each molecule of ethanol contains two carbon atoms, six hydrogen atoms, and one oxygen atom in a specific arrangement.

Matter can be described by its physical and chemical properties. Physical properties are those that can be observed without changing the identity of the matter. The color or luster of a metal is simply observed by sight. The density or temperature of a compound can be measured without changing its identity. On the other hand, the reactivity of matter can only be observed by performing a reaction. Examples of chemical properties include lithium metal's reactivity to water, the flammability of methane, and the voltage of a battery.

It is also important to be able to distinguish between physical and chemical changes in matter. Again, the difference is that physical changes do not alter the identity of the matter, but chemical changes do. Physical changes include melting, freezing, cutting, or dissolving. Chemical changes always involve a reaction and are often identified by a change in color or smell, or the formation of a solid or gas.

A.2   SI units

Chemistry Concepts: SI base units, SI derived units
Expected Learning Outcomes:
- Learn the SI base units
- Learn the SI prefixes
- Combine SI prefixes and base units to describe common measurements
- Manipulate SI base units to obtain SI derived units

The International System of Units (abbreviated SI) provides scientists around the world with an internationally agreed upon system for reporting measurements. While it is not the only system of measurement used in science and engineering, it has become increasingly accepted for use in other fields such as business and industry. Understanding SI units is important in many fields of work, and you probably

Table A.1 SI Base Units

Measure	SI Unit Name	SI Unit Symbol
Length	meter	m
Mass	kilogram	kg
Time	second	s
Electric current	ampere	A
Temperature	kelvin	K
Amount of substance	mole	mol
Luminous intensity	candela	cd

already know several units, even if you do not recognize them as being part of SI.

There are seven SI base units (Table A.1) that can be used alone or combined to form derived units. For example, acceleration in m/s^2 is a derived unit because it contains the SI base unit for length (meters) and the SI base unit for time (seconds) combined as meters per second squared. Some derived units have their own unique unit symbols. For example, s^{-1} is also known as hertz (Hz) and $kg\ m^2\ s^{-2}$ is also known as joules (J). To further illustrate this point, let us consider the determination of kinetic energy (in J), which is given by Equation (A.1).

$$KE = \frac{1}{2}mv^2 \tag{A.1}$$

The SI unit for mass is kg, and velocity is given in m/s (a derived SI unit). The units then combine as follows in Equation (A.2).

$$J = \frac{kg\ m^2}{s^2} \tag{A.2}$$

The prefixes shown in Table A.2 are used with SI units to show the magnitude of the units by factors of 10. A nanosecond (ns) is 10^{-9} s, or 0.000000001 s. A kilometer (km) is 10^3 m, or 1000 m. The kilogram (kg) is the only SI base unit that already contains a prefix. In the case of mass measurements, it is acceptable to report other prefixes with the unit grams (g). Thus, 10^{-9} kg becomes 1 µg. A more detailed explanation follows in Section A.4.

There are other units that are acceptable to use in conjunction with SI units, either because they are commonly used or easily derived from SI base units. Examples of these include the minute (min, 1 min = 60 s) and liter (L, 1 L = 1 dm³).

Table A.2 SI Prefixes

Greater than one			Less than one		
Factor	Prefix	Symbol	Factor	Prefix	Symbol
10^1	deka	da	10^{-1}	deci	d
10^2	hector	h	10^{-2}	centi	c
10^3	kilo	k	10^{-3}	milli	m
10^6	mega	M	10^{-6}	micro	μ
10^9	giga	G	10^{-9}	nano	n
10^{12}	tera	T	10^{-12}	pico	p
10^{15}	peta	P	10^{-15}	femto	f
10^{18}	exa	E	10^{-18}	atto	a
10^{21}	zetta	Z	10^{-21}	zepto	z
10^{24}	yotta	Y	10^{-24}	yocto	y

A.3   Scientific notation and significant figures

Chemistry Concepts: scientific notation, precision, accuracy, significant figures

Expected Learning Outcomes:

- Use scientific notation to handle large and small numbers
- Understand the difference between precision and accuracy
- Understand that significant figures allow us to report values with an idea of precision
- Learn the importance of checking values for sensibility

Scientists use a variety of means to express data and results in a manner that is convenient and conveys subtle information about the measurements. Sometimes very large or very small numbers must be condensed for convenience, and calculated numbers must be shortened to reflect the precision of the instruments used in collecting raw data.

Scientific notation is the standard way to express very large or very small numbers by shortening them to an abbreviated number and magnitude. The general form for scientific notation is given by Equation (A.3), where the magnitude N is adjusted such that $1 < N < 10$ and n is an integer, positive or negative, that gives the magnitude of the number.

$$N \times 10^n \tag{A.3}$$

The power of 10 essentially counts how many digits the decimal place moves from the original number to N. Thus, the number 341,000,000 becomes 3.41×10^8, and the number 0.0078 becomes 7.8×10^{-3}. A positive value of n expresses a number with an absolute value less than one.

Keep this point in mind when checking unit conversions or answers for sensibility.

There are two qualities of data that are often considered during collection and analysis: accuracy and precision. First, let us discuss accuracy, or the closeness of a measured number to its true value. If you were asked to draw a line 1 cm long without using a ruler, you might draw it accurately or you might overestimate or underestimate its length. Sometimes a measuring device or instrument is not properly calibrated and may do the same thing, yielding inaccurate measurements. The second quality to consider is precision, or the closeness of a number of measured values to one another. In this case, if you were asked to draw 10 lines that were each 1 cm long, you might draw them precisely of the same length, or if we look very closely, you might vary the lengths. In a similar manner, some measuring devices or instruments provide similar results for a measurement over and over again, while others vary considerably.

It is important to remember that precision and accuracy are not dependent upon one another. It is possible to have high accuracy and high precision (drawing 10 lines that are each very close to 1 cm), high accuracy and low precision (drawing 10 lines that are near 1 cm, but vary above and below), low accuracy and high precision (drawing 10 lines that are each very close to 1.3 cm instead of 1 cm), or low accuracy and low precision (drawing 10 lines that vary significantly in length above and below 1 cm).

A common analogy to describe accuracy and precision involves the game of darts (Figure A.1), where it is possible to have a series of throws result in high or low precision and high or low accuracy. Hitting the bull's-eye with five darts in a row would be considered to be both highly precise and accurate, hitting another section of the board with five darts would be precise but not accurate, getting five darts combined in and near the bull's-eye would be accurate but not precise, and randomly hitting the board would be neither accurate nor precise.

Accuracy can be reported in terms of a percent deviation from a known value, but precision is often reported subtly by the number of digits in a result. The digits reported are called *significant figures* (or sig figs), and the more significant digits a number has, the more precise the measurement is considered to be. There are a few simple rules for determining the number of significant figures in a number:

A. All nonzero digits are significant.
B. Zeros between significant digits are significant.
C. Zeros to the left of significant digits are not significant.
D. Zeros to the right of significant digits are significant if there is a decimal place in the number.

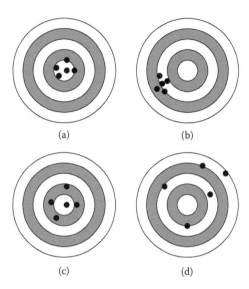

(a)   (b)

(c)   (d)

Figure A.1 Using a game of darts to visualize accuracy and precision, (a) represents high accuracy and high precision, (b) represents low accuracy and high precision, (c) represents high accuracy and low precision, and (d) represents low accuracy and low precision.

Let's consider a few examples in Table A.3 to understand these rules.

Scientific notation can be useful in reporting significant figures as well. Take, for example, the number 1400 in Table A.3. In scientific notation, it would be expressed as 1.4×10^3, which also has two significant figures. The number 0.00270 would be expressed as 2.70×10^{-3} because they both contain three significant figures.

Significant figures are carried through calculations to indicate the precision of the result based on the original data. Consider a calculator. Many calculators automatically give results with 8 to 10 digits when you input a simple calculation. If you divide 23.1 by 7.9, you get 2.92405063, but such a long number does not reflect the precision of the original data, in which there were three and two significant figures, respectively. It does not make sense to keep nine digits in the final answer, as that would imply a very precise result from imprecise data.

Two simple rules govern the determination of significant figures in calculated results. First, when multiplying or dividing, the number of significant digits in the result is the same as the lowest number of significant digits in the calculation. For the previous example, $23.1 \div 7.9 = 2.9$. The result is rounded to two significant digits because the calculation contained numbers with three and two significant digits, and the answer must be rounded to match the lower number of significant digits. For the product of $4.57 \times 9007 \times 0.000046$, the result requires two significant figures because

Table A.3 Examples of Significant-Digit Rules

Example	Number of significant digits	Rule(s) required
4.26	3	A
509	3	A, B
30.76	4	A, B
0.0048	2	A, C
0.024075	5	A, B, C
1400	2	A, D
6080	3	A, B, D
980.0	4	A, D
0.00270	3	A, C, D
0.0840500	6	A, B, C, D

the lowest number of significant digits comes from 0.000046 (two significant figures). The second rule is that when adding or subtracting, significant digits are determined by the number with the fewest decimal places. When performing these calculations, it often helps to write them down with the decimal places aligned, as seen in Equation (A.4).

$$
\begin{array}{r}
45.714 \\
+ \ 2.1 \\
\hline
47.8
\end{array}
$$

(A.4)

Note that the number of significant digits in the answer does not depend on the number of significant digits in either original number. The answer is rounded to the lowest number of decimal places.

One notable exception to these rules relates to pure numbers, or numbers that are exactly defined rather than measured. For example, if you count a dozen bolts, you know you have exactly 12 bolts and not potentially 12.31 or 11.84 bolts, since bolts only come in whole quantities. In this case, the number 12 is pure and will not affect the significant digits in any calculation (i.e., it will not be restricted to two significant figures unless another measured quantity in your calculation only has two significant figures). Similarly, exact unit conversions are pure numbers. Examples include 1 in = 2.54 cm and 1 m = 100 cm.

When you have finished any calculation, it is important to review your answer to see if it makes sense. For example, one common error is miswriting conversion factors and thus confusing a need to multiply or divide. If you were converting a wavelength of 653 nm to m and came up with the result 6.53×10^{11} m, a simple check would show that this answer does not make sense. The original wavelength of 653 nm should be less than 1 m because the nanometer is a very small unit (1 nm = 10^{-9} m), and this answer is far larger (in fact, it is on the order of the distance from Jupiter to the Sun!). The correct answer would be 6.53×10^{-7} m, a number far less than 1 meter.

A.4   Introduction to dimensional analysis

Chemistry Concepts: unit conversion, dimensional analysis, problem-solving skills, fundamentals of stoichiometry
Expected Learning Outcomes:
- Use dimensional analysis to convert between magnitudes of an SI base unit
- Use dimensional analysis to perform calculations using SI derived units
- Use dimensional analysis to perform simple stoichiometry calculations

Dimensional analysis is a method of converting between units, converting between unit magnitudes, or performing any number of other calculations in which units must be manipulated. In this section, we discuss dimensional analysis as a problem-solving tool.

There are two basic ideas to keep in mind when performing dimensional analysis. First, any equality can be written as a fraction equal to 1. We know that 1 in = 2.54 cm, so its conversion factor can be written as Equation (A.5)

$$\frac{1\ in}{2.54\ cm} \tag{A.5}$$

or Equation (A.6).

$$\frac{2.54\ cm}{1\ in} \tag{A.6}$$

Either of these fractions is equal to 1. The second notion is that to convert from one unit to another, the original unit must cancel with the unit in the conversion factor. If we wanted to convert 17.5 cm to inches, we would set up our calculation in such a way that centimeters become the bottom unit of the conversion factor in Equation (A.7).

$$17.5\ cm \times \frac{1\ in}{2.54\ cm} = 6.89\ in \tag{A.7}$$

In this manner, the units of centimeters cancel out just like variables in algebra, leaving us with an answer in inches. Note that only the *unit* cancels, not the number itself. As we discussed in the previous section, it is prudent to check the answer for sensibility. In this case, we know that inches are larger than centimeters, so the answer in inches should be less than the measurement in centimeters.

In Section A.2, we discussed SI units and ended with an explanation of how to treat the kilogram as a base SI unit when other prefixes are needed. The unit gram (g) is used with other prefixes, rather than combining multiple prefixes. This can be shown by the following Equation (A.8). Note that the units kilograms and grams cancel out, leaving micrograms.

$$10^{-9} \, kg \times \frac{10^3 \, g}{1 \, kg} \times \frac{1 \, \mu g}{10^{-6} \, g} = 1 \, \mu g \tag{A.8}$$

Squared or cubed units, such as those for area or volume, require special attention during calculations. It may be necessary to use several identical conversion factors to complete the calculation. If, for instance, we wanted to convert 1 cm³ to m³, we would have to do the following Equation (A.9).

$$1 \, cm^3 \times \frac{10^{-2} \, m}{1 \, cm} \times \frac{10^{-2} \, m}{1 \, cm} \times \frac{10^{-2} \, m}{1 \, cm} = 10^{-6} \, m^3 \tag{A.9}$$

Note that 10^{-6} m³ = 1 cm³, and the factor cannot be replaced by the prefix micro- (μ). 1 μm³ = 10^{-18} m³ because prefixes must cube when the units are cubed.

Knowing how to properly set up a dimensional analysis problem means that it is not necessary to memorize when to divide or multiply to convert between units or in other calculations, because setting up the units properly will yield the proper result. For example, if you needed to calculate how far a car could travel in 40 minutes at a speed of 96 km/hr, you would need to remember that 1 hr = 60 min and use dimensional analysis to cancel out all units but kilometers, as seen in Equation (A.10).

$$\frac{96 \, km}{hr} \times \frac{1 \, hr}{60 \, min} \times 40 \, min = 64 \, km \tag{A.10}$$

Dimensional analysis is used extensively in chemistry as part of solving stoichiometry problems, stoichiometry being the study of the quantitative amounts of substances in reactions. This book contains several examples of stoichiometry, and we will review the basics of these calculations here.

Remember that atoms are very small. It is not practical to count them in the quantities that we would normally see in a chemical reaction. Just as we count eggs or donuts by the dozen, we count atoms or molecules by the mole. A mole is 6.022×10^{23} units of a substance. So 1 mole of gold contains 6.022×10^{23} gold atoms and 1 mole of carbon dioxide contains 6.022×10^{23} molecules of carbon dioxide. Stoichiometry calculations in this book use molar mass to convert between grams and moles of a substance or use the mole to convert between two chemicals in a balanced reaction.

While moles are a more practical means of measuring amounts of chemicals than atoms or molecules because these entities are so small, we do not measure moles directly. Instead, we can measure the mass of a substance and convert it to moles using the molar mass of the substance. Molar mass is the mass equivalent of 1 mole of a substance. For atoms of an element, the molar mass in g/mol is equivalent to the atomic mass in amu. For compounds, the molar mass is determined by summing the molar masses of each atom in a compound. The atomic mass of Fe is 55.85 amu, so 1 mole of iron atoms has a mass of 55.85 g. Ethanol (C_2H_5OH) contains two carbon atoms, six hydrogen atoms, and one oxygen atom, so its molar mass is given by Equation (A.11).

$$(2 \times 12.01 \ g/mol) + (6 \times 1.008 \ g/mol) + (1 \times 16.00 \ g/mol) = 46.07 \ g/mol \quad \text{(A.11)}$$

One mole of ethanol has a mass of 46.07 grams.

Molar mass is used to convert between grams and moles. For example, to determine how many moles are in 70.4 grams of ethanol, we would set the calculation up as follows in Equation (A.12).

$$70.4 \ g \ C_2H_5OH \times \frac{1 \ mol \ C_2H_5OH}{46.07 \ g \ C_2H_5OH} = 1.53 \ mol \ C_2H_5OH \quad \text{(A.12)}$$

To convert from 0.743 moles ethanol to grams, we would set the calculation up as follows in Equation (A.13):

$$0.743 \ mol \ C_2H_5OH \times \frac{46.07 \ g \ C_2H_5OH}{1 \ mol \ C_2H_5OH} = 34.2 \ g \ C_2H_5OH \quad \text{(A.13)}$$

Note that in these calculations, we have started to include both the unit and chemical with each number, and both must cancel together. This is good practice, as you will see in the following examples.

A balanced chemical reaction gives the molar ratios of chemicals with their stoichiometric coefficients, which are the numbers in front of each chemical. The stoichiometric coefficients represent equivalents necessary to conduct a reaction or the equivalents obtained by performing the chemical process, much like the amounts and units of measurement in a recipe tell you how to prepare a dish. The combustion of propane is shown by the balanced reaction in Equation (A.14).

$$C_3H_8 + 5O_2 \rightarrow 3CO_2 + 4H_2O \qquad (A.14)$$

For every 1 mole of propane (C_3H_8) that combusts, 5 moles of oxygen (O_2) are consumed, 3 moles of carbon dioxide (CO_2) are produced, and 4 moles of water (H_2O) are produced. These mole ratios can be used in dimensional analysis calculations, just as any equality can. We can determine how many moles of carbon dioxide are produced from the combustion of 0.483 moles of propane by using the coefficients from the balanced reaction to show the mole ratio, as seen in Equation (A.15). Note that moles of propane cancel out, leaving moles of carbon dioxide. If we only include the unit (mol), but not the chemicals in this calculation, it would be easy to make a mistake.

$$0.483 \; mol \; C_3H_8 \times \frac{3 \; mol \; CO_2}{1 \; mol \; C_3H_8} = 1.45 \; mol \; CO_2 \qquad (A.15)$$

By combining the calculations involving molar mass and mole ratios, it is possible to answer even more complex stoichiometry problems. If we want to determine what mass of water can be produced from the combustion of 1.00 g propane, we would need to convert from grams of propane to moles of propane using the propane molar mass, from moles of propane to moles of water using the mole ratios from the balanced chemical reaction, and then from moles of water to grams of water using the molar mass of water, as shown in Equation (A.16):

$$1.00 \; g \; C_3H_8 \times \frac{1 \; mol \; C_3H_8}{44.1 \; g \; C_3H_8} \times \frac{4 \; mol \; H_2O}{1 \; mol \; C_3 \; H_8} \times \frac{18.02 \; g \; H_2O}{1 \; mol \; H_2O} = 1.63 \; g \; H_2O \quad (A.16)$$

In a stoichiometry problem, it is important to remember that molar mass is used to convert between mass and moles of a single chemical, and that the mole ratio is used to convert between moles of different chemicals.

Appendix B: Atoms and elements

B.1   Atomic structure

Chemistry Concepts: subatomic particles, electron configuration
Expected Learning Outcomes:
- Name the three fundamental subatomic particles
- Define *valence electron*
- Understand that electrons can be described with quantum numbers

Our understanding of the atom and its structure has changed greatly over time, with notable contributions by scientists during the atomic revolution of the early twentieth century. The evolution of our understanding of matter—from the ancients' classical elements to alchemy to quantum mechanics—has been an exciting journey of discovery spanning more than a thousand years. Today we recognize three subatomic particles that determine the structure and properties of atoms. Protons are positively charged and relatively heavy; neutrons have a neutral charge and are slightly heavier than protons; and electrons are very light, with a permanent negative charge. While some of these particles are made of even smaller particles (called *quarks*), the protons, neutrons, and electrons in an atom provide a sufficient basis for explaining chemical behavior. To fully comprehend reactivity, bonding, and light–matter interactions, it will be necessary to develop a basic understanding of electrons and their arrangement in the atom.

Clever experiments at the start of the twentieth century and the evolving theory of quantum mechanics taught us that the protons and neutrons are collected in a dense, positively charged core called a *nucleus*, while the electrons are constantly moving around the nucleus in a discrete number of relatively well-defined regions of space called *orbitals*. In terms of atoms, atomic orbitals are arranged into shells and subshells that help us to keep track of the energy and momentum associated with electrons in each orbital. Each orbital can only hold a maximum of two electrons, and therefore each shell with a discrete number of orbitals can only hold a certain number of electrons. Shells that have their full complement of electrons are considered to be *core* shells, and though they affect atomic properties, they are not directly responsible for chemical behavior. Shells that have fewer than their full complement of electrons are called *valence* shells, and it is the electrons in these valence shells/valence orbitals that are responsible for the chemical and photochemical reactivity of atoms and molecules. For example, when atoms combine to form molecules via covalent bonding, the outermost electrons are shared by the species involved and reside in new orbitals with shapes that differ from those of the individual atomic orbitals (see Chapter 7).

Since positive and negative charges attract one another (electrostatic attraction) and like charges repel one another (electrostatic repulsion), the energy that a particular electron requires to stay in its orbital is very closely related to how far each electron is, on average, with respect to the positively charged nuclei and other negatively charged electrons in an atom or molecule. Because every orbital in an atom or molecule encompasses different regions of space, each electron has a unique set of distances with respect to the other electrons and nuclei, and therefore electrons in different orbitals exhibit different energies. Likewise, since the number and arrangement of protons and electrons in each atom or molecule is different, every atom and molecule has a unique set of orbitals and unique orbital energies specific to that atom or molecule. This unique set of orbital energies is the essential foundation of light spectroscopy and is also crucial to understanding color (Chapter 7).

Atoms and molecules by nature will adopt the lowest energy configuration of electrons, which we call the *ground state* electron configuration. The ground state configuration is generated by filling the available orbitals in an atom with electrons, starting at the lowest energy orbital and moving to the next highest energy orbital until all the electrons have been assigned by using the Aufbau principle and by following Hund's rule, which says that only two electrons can occupy a single orbital, provided that they have opposite spin angular momentum. In the most basic sense, we can consider the electrons to occupy orbitals much like people occupy rows of seats at a concert. Just as fans want to be as close to the stage as possible, electrons will fill in the spaces closest to the nucleus first to minimize the energy required to keep an electron from crashing into the nucleus under the influence of electrostatic attraction. Fans have tickets that designate their unique sections, rows, and seat numbers; electrons are assigned a set of quantum numbers designating their shell, subshell, orbital orientation, and spin. Each electron in a given atom will have a unique combination of these four numbers, ensuring that each electron is unique.

Atoms always have more orbitals available than there are electrons, so there are empty orbitals that are not occupied in the ground state where an electron can establish a new stable position with respect to the nucleus. Because of the Aufbau principle, the first empty orbital in an atom or molecule will always be found at higher energy than the last electron in a ground state electron configuration. One can loosely think of the discrete orbitals discussed here as rungs on a ladder or steps on a staircase. When you climb one of these devices, you can stop on any rung or stair, but you cannot stop in between. To move up a stair or rung, you must also put enough energy into lifting your leg that it reaches or goes above the level of the next stair or rung. Likewise, an electron must receive an amount of energy equal to or greater than the gap between its current and final orbital if it is to make a transition to a new stable orbital at one of the allowed values of energy. When an electron eventually hops from this higher energy "excited state" back to the ground state position, it must give up a fixed amount of energy equal to the gap between the orbitals, much like you give up a fixed amount of gravitational potential energy when you take a step down a staircase.

B.2   Introduction to the periodic table

Chemistry Concepts: periodic table
Expected Learning Outcomes:
- Explain the history of the periodic table
- Label the major group names on the periodic table
- Look up information in the periodic table

There are many resources available that detail the discovery of the elements that occupy the modern periodic table, so we will not dwell on the history of elemental discovery. However, you should be aware that the oldest known elements are metals, such as copper and gold, that can be found in their native states in nature. Modern scientific discovery of elements did not begin until the seventeenth century, and the periodic table was not published in its first form until 1869, when Dmitri Mendeleev organized the 63 known elements at the time by mass and properties. Mendeleev's table predicted the existence of several unknown elements based on gaps in the table (these elements were discovered later).

The modern periodic table, as shown in Figure B.1, contains well over a hundred elements in order of atomic number (the number of protons) and organized by electron configuration and common chemical properties. Columns on the periodic table are called groups, while rows are called periods. Groups are given numerical designations as well as names. Common group names are given in Table B.1.

In addition to group names, portions of the periodic table are referred to as *blocks*. These blocks designate information about the location of valence electrons in the atom. The s-block includes groups 1 and 2, or the

1																	18
1 H	2											13	14	15	16	17	2 He
3 Li	4 Be											5 B	6 C	7 N	8 O	9 F	10 Ne
11 Na	12 Mg	3	4	5	6	7	8	9	10	11	12	13 Al	14 Si	15 P	16 S	17 Cl	18 Ar
19 K	20 Ca	21 Sc	22 Ti	23 V	24 Cr	25 Mn	26 Fe	27 Co	28 Ni	29 Cu	30 Zn	31 Ga	32 Ge	33 As	34 Se	35 Br	36 Kr
37 Rb	38 Sr	39 Y	40 Zr	41 Nb	42 Mo	43 Tc	44 Ru	45 Rh	46 Pd	47 Ag	48 Cd	49 In	50 Sn	51 Sb	52 Te	53 I	54 Xe
55 Cs	56 Ba		72 Hf	73 Ta	74 W	75 Re	76 Os	77 Ir	78 Pt	79 Au	80 Hg	81 Tl	82 Pb	83 Bi	84 Po	85 At	86 Rn
87 Fr	88 Ra		104 Rf	105 Db	106 Sg	107 Bh	108 Hs	109 Mt	110 Ds	111 Rg	112 Cn		114 Fl		116 Lv		

57 La	58 Ce	59 Pr	60 Nd	61 Pm	62 Sm	63 Eu	64 Gd	65 Tb	66 Dy	67 Ho	68 Er	69 Tm	70 Yb	71 Lu
89 Ac	90 Th	91 Pa	92 U	93 Np	94 Pu	95 Am	96 Cm	97 Bk	98 Cf	99 Es	100 Fm	101 Md	102 No	103 Lr

Figure B.1 The modern periodic table contains well over 100 known and named elements organized in order of atomic number (number of protons) and electron configuration. Each column, or group, is numbered.

Table B.1 Common Group Names of the Periodic Table

Group number	Group name	Elements
1	Alkali metals	Li, Na, K, etc.
2	Alkaline earth metals	Mg, Ca, Sr, etc.
17	Halogens	F, Cl, Br, etc.
18	Noble gases	He, Ne, Ar, etc.

alkali and alkaline earth metals. The p-block includes the elements boron through neon and all elements below them that reside in the same columns. The d-block, or transition metals, includes the elements scandium through zinc and all elements below them in those columns. The f-block, at the bottom of the periodic table, includes the remaining elements (often referred to as the *lanthanoids* and *actinoids*).

There are many versions of the periodic table, each with a different format and varying types of information. Many periodic tables will provide the atomic number, symbol, and atomic mass for each element. In general, the atomic number (1 for hydrogen, 2 for helium, 3 for beryllium, and so on) will be above the element's symbol. Each element has a one- or two-letter symbol (H for hydrogen, He for helium, Be for beryllium, and so on). While the names of elements and chemicals are not capitalized, the first letter of the symbol of each element is capitalized. In many periodic tables, the atomic mass is provided below the element symbol. Atomic mass is the average mass of each atom in 1 mole, or 6.022×10^{23}

atoms, of that element and is equivalent to the average mass in grams of 1 mole of that element. More detailed versions of the periodic table may provide more specific and specialized information about each element, but such editions will generally provide a legend detailing how to decipher each entry.

B.3   Some periodic trends

Chemistry Concepts: valence, ions, atomic size, metallic properties, reactivity

Expected Learning Outcomes:

- Describe how trends in valence affect:
  - Reactivity
  - Ion charge
  - Atomic size
- Describe how metallic properties change across the periodic table

Recalling the seat analogy of electronic structure from Section B.1, we can think of the periodic table as a layout of seats at a concert. The first row only contains two seats, and the second and third rows each contain eight seats, somewhat like stadium seating, where the number of seats per row increases from the inner to the outer rows in a section. An atom of phosphorus contains 15 electrons, so they will fill in the entire first and second rows, with five electrons left to occupy the third row. In an atom, we consider these concert rows to be shells with increasing mean (or average) distance from the nucleus. Thus, the third shell of phosphorus contains five electrons, which we call valence electrons. The lower (or core) shells are filled, stable, and unreactive, while the outer shell contains the valence electrons that are free to react with other atoms with the goal of filling the outermost shell completely by gaining, losing, or sharing electrons.

The number of valence electrons in an atom follows a general trend across the periodic table. Group 1 elements will have a single valence electron, and group 2 elements will have two. Then we jump to group 13 with three valence electrons, and the trend continues across the periodic table to group 18 with eight valence electrons.

Group 18 elements, the noble gases, are generally unreactive, as their outermost shell is already filled. As you might imagine, elements that only need to gain or lose a single electron are very reactive. Group 1 elements, or the alkali metals, only have a single valence electron. Losing this electron yields a +1 charge. The alkali metals are known for their violent reactions with water, and reactivity increases as one moves down this group in the periodic table. As more inner electron shells are filled, the single valence electron is farther from the nucleus and shielded from its positive charge by core electrons, making it even easier to lose. Similarly, group 17 elements, the halogens, have seven valence electrons and need only gain one electron (to obtain a −1 charge) to reach a stable, filled-shell electron configuration. Since the halogens gain electrons to obtain a full outer shell, they become less reactive down the periodic table. Inner shells that are more filled shield the electron being added to the atom from the positive charge on the nucleus, making it less attracted. Thus, fluorine is the most reactive of the halogens.

In this discussion of reactivity, we mentioned the formation of +1 charged cations in alkali metals and −1 charged anions in halogens. As you may guess, alkaline earth metals (those elements in group 2) can lose their two valence electrons to form ions with a +2 charge. Charges below period three in the d- and p-blocks are more difficult to predict due to complexities that arise with greater numbers of electrons, but it is easy to predict charges of ions such as O^{2-}, Al^{3+}, and N^{3-} because they are higher on the periodic table and have a small number of core electrons and a limited number of "seats" in their electron stadiums.

From our discussion of electron organization in atoms, it is apparent that as the number of core electrons contained in an atom increases, the outermost electrons are farther from the nucleus. As you go down the periodic table in any given group, atomic size increases due to the larger number of electrons occupying more core shells. However, as you go across a given row in the periodic table, atomic size actually decreases. This is because the number of core electrons remains constant, shielding the positive charge from the nucleus in a similar manner. Meanwhile, the nucleus is gaining protons as we move across a row in the table. Thus, the effective positive charge experienced by the valence electrons increases across a given row in the periodic table, and that greater positive charge provides a greater attraction to the valence electrons. This attraction, in turn, makes each element decrease in size across a given row.

In the study of automobiles, the characteristics of metals are exceedingly important. In the periodic table, metallic character decreases across a period (from left to right) and increases down a group. Most elements on the periodic table are indeed metals, with only a triangle on the upper right side of the periodic table belonging to nonmetals (imagine a line from C to Rn, and everything including these and above and to the right are nonmetals). Nonmetals provide us with the complex chemistry that makes oils, plastics, and polymers, while the characteristics of metals yield strong engine blocks, conductive wires, and shiny chrome.

We are accustomed to talking about the properties of metals in several terms. *Malleability* is the ability of a material to be hammered or rolled via compression. *Ductility* is the ability of a material to be formed into a wire via tensile stress. While malleable and ductile metals overlap to some extent, they are not mutually inclusive due to the different types of stress required to achieve these physical changes. Most metals have shiny luster, or a surface that reflects light. However, some of the softer metals such as lithium and calcium oxidize readily, and this luster can only be seen when they are cut under an inert atmosphere or mineral oil. Another property common to many metals is the ability to conduct heat or electricity. Silver, copper, and gold have the highest electrical conductivities among pure metals.

Appendix C: Organic chemistry

C.1 Nomenclature

Chemistry Concepts: organic nomenclature, functional groups, line drawings

Expected Learning Outcomes:
- Name simple organic molecules
- Recognize functional groups
- Convert between organic structures and line drawings

The field of organic chemistry encompasses the chemical arrangements and rearrangements of carbon and other nonmetals to form myriad combinations of molecules, ranging from simple methane (natural gas) containing five atoms to plastics, proteins, and DNA that may contain millions or billions of atoms. As with all science, it is important to have a standard and internationally agreed upon method for describing and naming these molecules so that we can communicate with one another and the general public effectively, particularly as the molecules become more complex. The International Union of Pure and Applied Chemistry (IUPAC) is the body charged with recommending a nomenclature (naming system) for organic molecules to avoid confusion and facilitate international communication. While IUPAC guidelines are not always followed in cases where common names are widely known, it is helpful to have an understanding of this system of nomenclature. We discuss the basic aspects of the naming system here, but more complete and detailed naming discussions can be found in IUPAC literature[*] or in any organic chemistry textbook.

In organic molecules, atoms share electrons in covalent bonds in order to fill their outermost electron shells, thereby arriving at stable electron configurations for each of their constituent elements. Because each atom starts with a given number of valence electrons, elements tend to form specific numbers of bonds when incorporated into an organic molecule.

---

[*] R. Panico, W. H. Powell, and J-C. Richter, *A Guide to IUPAC Nomenclature of Organic Compounds Recommendations 1993* (Oxford, UK: Blackwell Scientific Publications, 1993).

The most common elements in organic compounds are carbon (C), hydrogen (H), oxygen (O), nitrogen (N), phosphorus (P), sulfur (S), and the halogens. Carbon will generally form four bonds, nitrogen and phosphorus three, oxygen and sulfur two, and hydrogen and halogens form one bond. This is because carbon requires four additional shared electrons to fill its valence shell, phosphorus and nitrogen require three, oxygen and sulfur two, and hydrogen and halogens one additional shared electron. When an atom can form more than one bond, its bonds can be a combination of single (two electrons shared with one partner), double (four electrons shared with one partner), or triple (six electrons shared with one partner) bonds. For example, carbon can gain the four electrons it needs for a complete valence shell by forming four single bonds, a single bond and a triple bond, a double bond and two single bonds, or two double bonds. These common bonding patterns yield predictable atomic arrangements in molecules. For example, because carbon can form four bonds, it will always be on the inside of the molecule, surrounded by other atoms. On the other hand, hydrogen and halogens are always on the outside of the molecule because they will only form a single bond.

With larger and more complex molecules, you might imagine that drawing every single atom might become tedious and make it difficult to render a molecular diagram that provides useful insight. Because carbon and hydrogen are by far the most abundant elements in organic molecules and we know what their bonding patterns are, we often do not draw them explicitly when we draw pictures of molecular structures. Instead, we use what is called a *line drawing* or *skeletal structure*, where carbon is assumed to occur at each line terminus or intersection, and hydrogen atoms are assumed to be present as needed to fulfill carbon's need for four bonds on each atom. Every atom that is not carbon or hydrogen is shown explicitly, as are hydrogen atoms that are not attached directly to a carbon atom. For example, octane can be drawn identically as shown in Figure C.1, with all atoms shown on the left drawing and the simpler line drawing on the right. Likewise, ethylene glycol could be represented as shown in

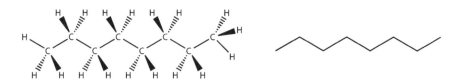

Figure C.1 Structural diagrams of octane. The explicit model on the left and line drawing on the right are equivalent. In these structural drawings, some three dimensionality is depicted with solid wedges (coming out of the plane of the paper toward you) and dashed wedges (going out of the plane of the paper away from you).

Figure C.2 Detailed (left) and line (right) structures of ethylene glycol.

Figure C.2, with all atoms being shown on the left drawing and the simpler line drawing on the right. Throughout the rest of this appendix, we use line drawings when practical.

According to IUPAC recommendations, organic molecules are described by defining their backbone, side chains, and functional groups. The backbone is considered the longest chain of bonded carbon atoms in the molecule. Side chains are branches of carbon(s) that stick out from the backbone, like branches stick out from the trunk of a tree. Functional groups contain other atoms in addition to carbon and hydrogen and are so named because they are specific arrangements of atoms that impart distinctive characteristics and chemistry to a molecule.

The carbon backbone of an organic molecule is systematically named to describe how many carbon atoms are present and what kinds of bonds are present. A carbon chain that only contains single bonds is called an *alkane*. If any double bonds are present, it is an *alkene*. If any triple bonds are present, it is an *alkyne*. The suffixes *-ane*, *-ene*, or *-yne* will tell you the highest order carbon bond (the one with the most electrons shared between two adjacent carbon atoms) present in the molecule. For our purposes, we only discuss the proper naming of alkanes here. Prefixes in backbone names denote how many carbon atoms are present in the longest carbon chain. Table C.1 describes the prefixes and examples of alkanes.

When a carbon chain is branched (not linear), we first name the longest chain and then identify any branches with prefixes that identify the number of carbons in a given branch. For example, a branch with only a single carbon is called a methyl group, a branch with two carbon atoms is an ethyl group, and so on. In order to describe the location of the branch, the carbons on the backbone are numbered from end to end such that the backbone carbon atom where the branch occurs has the lowest number possible. For example, the molecule in Figure C.3 would be named 3-ethylhexane, meaning that it has a six-carbon backbone (hexane) with a branch containing two carbon atoms (ethyl) attached to the third carbon in the backbone (3-). Note that this molecule has the same chemical formula as octane, but a very different structure.

Functional groups can exist as branches attached to a carbon backbone or be located between two separate carbon chains. Each functional group in organic chemistry has the same arrangement of atoms anytime it

Table C.1 Naming Alkanes

Number of carbon atoms	Name	Structure
1	methane	CH_4 (H—C—H with H above and below)
2	ethane	CH_3—CH_3
3	propane	CH_3—CH_2—CH_3
4	butane	CH_3—CH_2—CH_2—CH_3
5	pentane	CH_3—CH_2—CH_2—CH_2—CH_3
6	hexane	(skeletal zig-zag structure)
7	heptane	(skeletal zig-zag structure)
8	octane	(skeletal zig-zag structure)
9	nonane	(skeletal zig-zag structure)
10	decane	(skeletal zig-zag structure)

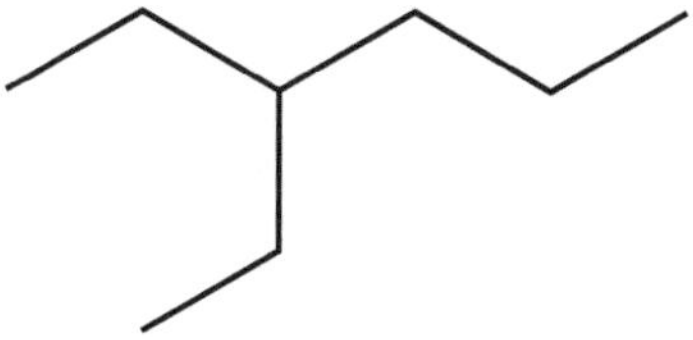

Figure C.3 Structure of 3-ethylhexane.

Table C.2 Common Functional Groups

Name	Prefix or suffix	Structure
Alcohol	-ol	−OH
Halogen	fluoro-, chloro-, bromo-, iodo-	−F, −Cl, −Br, −I
Ketone	-one	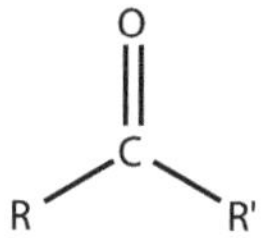
Aldehyde	-al	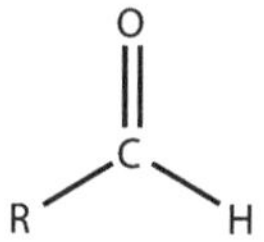
Carboxylic acid	-oic acid	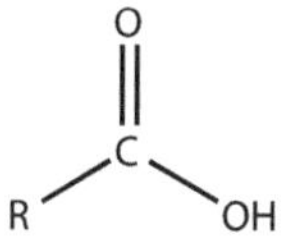
Ether		
Amine	-amine	R—NH$_2$

appears and can participate in identical or similar reactions depending to some extent on what it is bonded to. A list of common functional groups is given in Table C.2. When naming an organic molecule with functional groups, we use a similar system to that described previously. The carbon atoms in the backbone carbon chain are numbered so that the smallest possible number is given to the carbon bonded to the functional group, and a prefix or suffix is used to identify the functional group. For example, Figure C.4 contains 2-pentanol on the left and 2,3-difluorohexane on the right. The name *2-pentanol* means that the molecule has a five-carbon backbone and an alcohol group on the second carbon. The 2,3-difluorohexane has a six-carbon backbone with a single fluorine atom bonded to each of two carbons, the second and third, in the backbone.

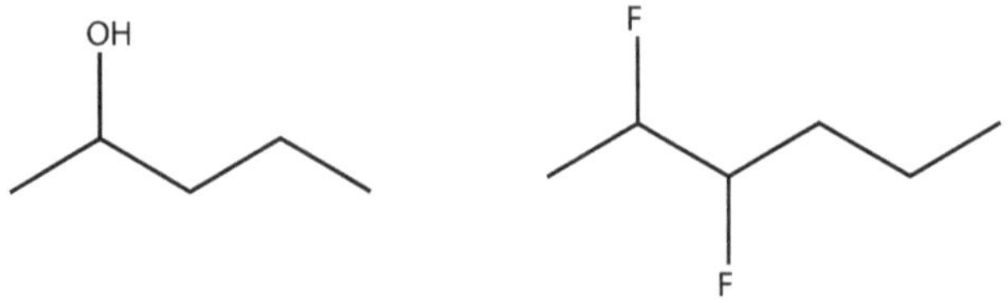

Figure C.4 Structures of 2-pentanol (left) and 2,3-difluorohexane (right).

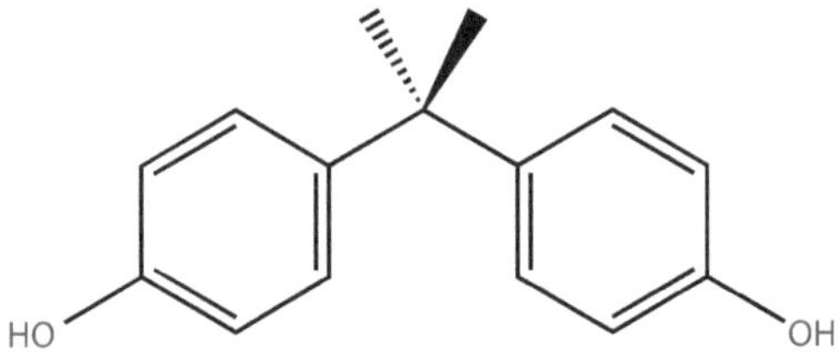

Figure C.5 Structure of bisphenol A.

C.2   Polymerization

Chemistry Concepts: organic reactions, polymers
Expected Learning Outcomes:
- Define monomer, polymer, and polymerization
- Explain the relationship between the structure and function of polymers
- Write a few simple polymerization reactions

One important subset of organic compounds in cars and other consumer products are the polymers. You come in contact with polymers every day in the form of plastics, rubber, synthetic fabrics, and/or silicone. In fact, your body depends on the natural polymers of DNA and proteins to function. Polymers are large organic molecules made of regularly repeating units. These smaller units are called *monomers* and typically correspond to smaller organic molecules that link to form the polymer. The process of combining parent monomer molecules together to form larger molecules is called *polymerization,* and the physical and chemical properties of the large polymers differ significantly from their parent monomers. In this section, we look at some typical examples of monomers and polymers used in cars, their characteristics, and some of the reactions that form them.

One monomer in particular that has garnered major media and health attention in the past few decades is bisphenol A (BPA), as seen in Figure C.5. BPA is used in the production of epoxy resins and polycarbonates that appear frequently in vehicle headlamps, windows, and plastic panels (interior and exterior), and has been used in products as diverse as baby bottles, cash register receipts, DVDs, and cans of food. However, in the body, BPA binds to estrogen receptors, which has the potential for serious health consequences. Because the results of medical studies vary regarding the actual effects of BPA on the human body and because BPA can be detected in most people, its use has become quite controversial, leading to bans and proposed bans in several countries (mainly to avoid its use in baby bottles and children's products) and many voluntary reformulations of plastics that come in contact with food.

Another important class of monomer is the *epoxide*, which is a general term for a molecule that contains a triangular atom arrangement (three-atom ring) consisting of two carbons and an oxygen atom bound together (a cyclic ether), with other atoms bound to the carbons. In the generic form depicted in Figure C.6, the groups designated R can be any other portion of an organic molecule, from a single hydrogen atom to a larger carbon chain with or without other functional groups.

This cyclic epoxide has a rather unstable atomic arrangement and reacts readily. The molecules in these epoxide resins in turn, can react with other monomers, they become epoxide resins. A resin is a thick flowing liquid that hardens into a clear solid. These epoxide resins, in turn, can react with one another to form long polymers, or they can be treated with a curing agent to cross-polymerize and form what is commonly known as an *epoxy*. Let's look at this process for the combination of BPA and epichlorohydrin (Figure C.7), an epoxide. When combined, they form an epoxy resin called bisphenol A diglycidyl ether (BADGE), depicted in Figure C.8, which can form a long, repeating diepoxy polymer that will then react with a curing agent such as ethylenediamine, shown in Figure C.9.

Each molecule of ethylenediamine can react with four molecules of the diepoxy, with each diepoxy reacting at one of the N–H bonds of the ethylenediamine. While we cover the mechanism in detail here, note that the hydrogen atoms that are displaced from ethylenediamine do go to the diepoxy molecule whether it is a single molecule of BADGE or a polymer, forming an –O–H bond where there was previously a =O bond.

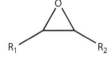

Figure C.6 Generic structure of epoxide.

Figure C.7 Structure of epichlorohydrin.

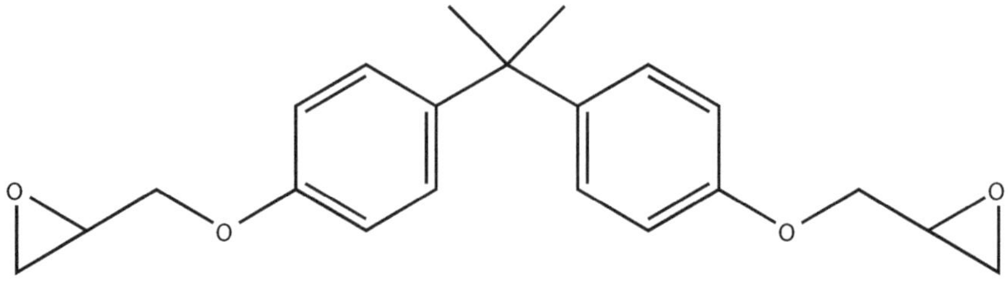

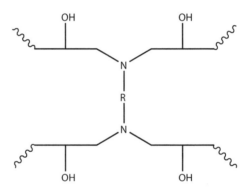

Figure C.8 Structure of bisphenol A diglycidyl ether (BADGE).

Figure C.9 Structure of ethylenediamine.

Figure C.10 An example of an ethylenediamine cross-linked polymer epoxy.

Because the molecule has reactive sites on both ends, this reaction can make a complex network that is one very large molecule.

In Figure C.10, only the cross-linking site is shown in detail. A cross-link is a chemical bond between two polymer chains that links them together. The wavy lines indicate the remainder of the diepoxy, whether it is a single molecule of BADGE or a large polymer derived from the BADGE monomer. In this example, the ethylenediamine is a cross-linking agent, meaning that it combines the smaller molecules in a liquid form into much larger molecules that form a solid with a large mean molar mass. Cross-linking alters the properties of a compound by making it less prone to plastic (irreversible) deformation and by increasing the crystallinity of the polymer, and may modify other physical properties.

Another polymer derived from BPA is polycarbonate, which is a strong and lightweight polymer used in compact discs, eyeglasses, headlamps and bumpers, and it can even be laminated in thick layers to form bullet-resistant windows. Polycarbonate is formed in an overall reaction of BPA with phosgene, as seen in Figure C.11. Note that only one unit of polycarbonate is actually shown as a product in this reaction. The brackets and subscript letter *n* denote that this structure repeats over and over again in the large polymer molecule.

Another polymer with which you are probably familiar is polyvinyl-chloride (PVC). It is formed from the monomer chloroethene, or vinyl chloride. Figure C.12 shows the structure of vinyl chloride monomer and PVC polymer. PVC tends to form strong, rigid plastic components with relatively low temperature stability. The properties of the final product can be altered by the addition of additives to change the rigidity, color, and chemical stability of the polymer. Because such an array of characteristics can be achieved, PVC is used in a wide range of applications including pipes, electrical insulation, and leather alternatives.

Polystyrene is another common polymer with automotive, consumer, and industrial applications. It is made from the monomer styrene. Figure C.13 shows the structures of styrene and polystyrene. As seen in the figure, polystyrene has long carbon chains with attached phenyl groups (six-membered rings of carbon with alternating double and single bonds). This polymer only contains carbon and hydrogen atoms and thus can only develop dispersion forces between molecules. It is unreactive

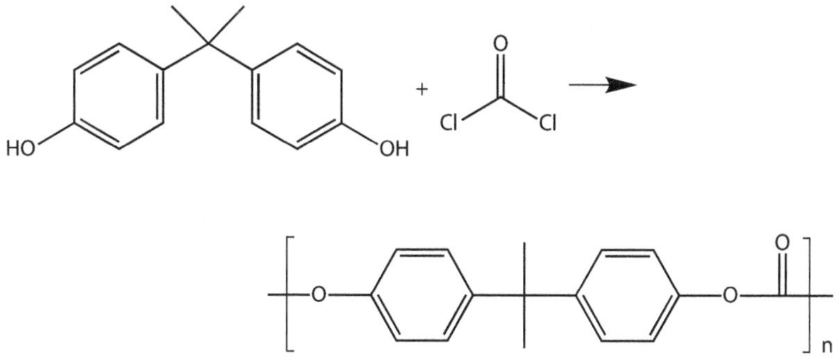

Figure C.11 Reaction of BPA and phosgene.

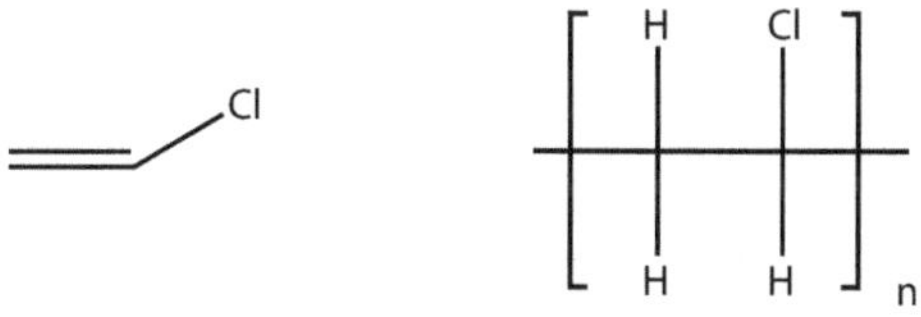

Figure C.12 Structure of vinyl chloride monomer and PVC polymer.

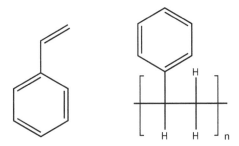

Figure C.13 Structures of styrene and polystyrene.

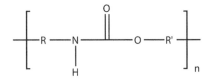

Figure C.14 Structure of polyurethane.

with most other chemicals. As a plastic, polystyrene is hard and brittle. It is used in applications where cheap and disposable plastic is desired, such as cutlery and disposable lab vessels. You are probably more familiar with its foam form, which is used in disposable cups for hot beverages, takeout containers, insulation, and automobile bumper cores.

Polyurethane is another common automotive polymer that also has both plastic and foam applications. Its monomer and formation chemistry is more complex than the previous polymer examples, but a generic polyurethane structure can be found in Figure C.14. Once again, the *n* denotes that this single unit is repeated over and over within a larger polymer molecule. The diverse polyurethanes that can be made with different R and R′ components means that this class of molecule can have a wide variety of physical and chemical properties. Elasticity and heat resistance are among the characteristics that can be altered by using different monomers. Polyurethane foams have extensive applications in upholstery forms and insulation. Other applications include skateboard wheels, bushings, Spandex, and varnish.

Polymers can be classified both by their composition and the reactions that form them. Homopolymers, like PVC and polystyrene, are those that are formed from a single monomer. Copolymers, like epoxies, are formed from a single monomer, Heteropolymers or copolymers, like epoxies, are formed from more than one type of monomer. The reactions that form polymers are considered to either be step growth or chain growth. The topics outlined here are discussed in greater detail as part of Chapter 6.

Appendix D: An introduction to the car

D.1 Major systems in the car

A car is a complex piece of machinery that is arguably the ultimate intermingling of art, engineering, and science. This complexity can make the task of learning about cars overwhelming to someone just starting to study cars or someone trying to become more familiar with cars and how they work. However, we can make the task more manageable (and fun) by taking advantage of "systems." In science, a system is the part of the universe we wish to study, and everything else is considered to be the surroundings. The entire car can be considered to be one system. However, we could also consider the master car system as containing many different integrated and interdependent subsystems that each handle one of the major functions of a car.

The advantage of subsystems is that they allow us to tackle smaller and more easily understandable pieces of the car until we understand the system as a whole. The good news is that we do not have to come up with definitions and an organization system of subsystems on our own; cars have been around long enough that many predefined subsystems incorporating the crucial components already exist. In this appendix, we will provide a list of the most important automotive subsystems and an overview of how several of these systems operate. This is by no means meant to be a complete primer to understanding a vehicle. The interested reader is highly encouraged to look up a term or system in another book or online for additional details or to find a variety of useful diagrams.

The major subsystems in a car include

- *Chassis*: The structural components of the car that give it strength and shape. This includes the frame and subframe in trucks and older cars or the unibody/monocoque in more modern vehicles.
- *Powertrain*: The subsystem that powers the vehicle, causing it to move. The power train starts with the power-generating component

and extends down to the wheels that receive power and push/pull the vehicle. The drivetrain itself is composed of several subsystems.

- *Motor*: The device that harvests power from a chemical reaction, be it a combustion-based engine or an electrochemically driven electric motor using current generated by electrochemical reactions in a battery or fuel cell.
- *Transmission*: The device that converts rotation of the motor or engine flywheel to power at the wheels, usually with several selectable gear ratios that help to make acceleration smooth and rapid.
- *Half shafts/driveshafts*: The components that carry the power from the transmission to either the wheels (in the case of half shafts) or to a differential.
- *Differential*: A component that splits the incoming power and shifts the orientation 90° so that it can reach the wheels. Applies only in rear-wheel-drive (RWD) and all-wheel-drive vehicle configurations (AWD/4WD).
- *Final drive*: The components that actually apply the power to the ground, including the wheels, wheel hubs, wheel bearings, etc.
- *Drivetrain*: The components of the power train other than the engine/motor.
- *Intake system*: The collection of parts that draws in air for the combustion process. This includes a pipe open to the outside air called the *air intake*, an air filter, a throttle body (a valve that controls how much air the engine is drawing in), and an intake manifold that distributes and channels the air to the intake valves.
- *Exhaust system*: The collection of parts that manage the waste exhaust gases from an internal combustion engine. This includes exhaust pipes, catalytic converters, and mufflers.
- *Fuel*: The components that store fuel and transmit it to the engine. This includes the gas tank, fuel pump, fuel lines, carburetor/fuel injectors, etc.
- *Cooling system*: The components that remove waste heat from the engine and help to maintain a constant operating temperature. This includes the radiator, water pump, radiator hoses, radiator fans, and the liquid coolant.
- *Suspension*: The components that attach the rest of the car to the wheels and permit motion of the rest of the vehicle relative to that of the wheels. In other words, it allows the wheels to remain in contact with the road and the passenger cabin to be semi-isolated from bumps, potholes, chassis roll, etc.
- *Steering system*: The parts of the car that allow the driver of a vehicle to control the direction in which the vehicle is moving. Includes the steering wheel, steering column, steering rack, power steering pump, power steering fluid, tie rod, and ball-bearing kingpin.

- *Electrical system*: The parts of the car responsible for storing and transmitting electricity throughout the vehicle. This includes the battery, alternator, ignition, starter motor, wiring, fuse boxes, many types of sensors, lights inside the vehicle, taillamps and headlamps, power door locks, power mirrors, rear window defroster, etc.
- *Ignition system*: The components that produce the spark in conventional gasoline internal combustion engines. This includes a charge-distribution mechanism, spark plug wires, and spark plugs.
- *Stereo system*: The components responsible for generating music. This includes the head unit, amplifiers, wiring, and speakers.
- *Air conditioning system*: The components that cool the interior of the vehicle. This includes the compressor, evaporator coils, pressure-drop valve, condenser coils, fans, cabin filters, and intermediate plumbing.
- *Supplemental restraint system*: The series of sensors and air bags that control air-bag deployment during an emergency.
- *Braking system*: The components that help to stop the car quickly. This includes the brake calipers, brake pads, brake lines, brake rotors, brake fluid, and brake pedals.
- *Emission control system*: A series of sensors and a computer that help to manage the fuel/air ratios to achieve an appropriate balance of power and low pollutant emissions. The system includes a number of sensors (mass airflow sensor, oxygen sensor, etc.), the computer, positive crankcase ventilation system, catalytic converter, etc.

D.2   Drivetrain configurations and components

Recall that the drivetrain is all the components of the powertrain other than the engine. There are several possible configurations for the drivetrain of your vehicle, including front-wheel drive (FWD), rear-wheel drive (RWD), all-wheel drive (AWD), and four-wheel drive (4WD). In front-wheel drive, the front wheels on the vehicle are the ones that receive the power and pull the car. FWD configurations include engines mounted in a transverse orientation (the long dimension of the engine, or the crankshaft, lie parallel to the axles) and the transmission turns half-shafts that send the power to each front wheel. Key to FWD are the constant velocity (CV) joints that allow power to get to the wheels while the rest of the drivetrain moves relative to the wheel via the suspension. The interested reader is highly encouraged to look up CV joints online for many graphic demonstrations of how they work. CV joints are also used in rear-wheel-drive cars that have fully independent rear suspensions, meaning both rear wheels travel independently from one another rather than being linked through an axle. RWD cars send the power from the engine to the rear wheels and have engines mounted longitudinally (with the crankshaft perpendicular to the axles). The transmission in RWD cars turns a driveshaft, which sends power to a rear differential. The differential contains a series of gears that allow the rear half-shafts to turn at the same rate as the driveshaft, effectively rotating the power 90° so that the wheels can drive the vehicle. AWD cars have much the same configuration as RWD, except that there are both front and rear differentials. The transmission sends power via a driveshaft to a central differential that rotates other driveshafts that carry power to the front wheels via the front differential and the rear wheels via the rear differential. 4WD is much the same as AWD, except instead of a central differential, power from the transmission goes to a transfer case that will only allow the front and rear driveshafts to turn at exactly the same rate, with a 50/50 power distribution between front and rear.

D.3   Engines and engine configurations

The engine in a combustion-based vehicle is the device that captures the chemical energy released in a combustion reaction and transforms a portion of it into the mechanical energy that drives the vehicle. All internal combustion engines contain essentially the same parts, and a detailed discussion of the engine operating cycle and internal components can be found in Chapter 2. These parts change very little whether you are considering conventional gasoline, diesel direct-injection, or gasoline direct-injection engines, other than the fact that the ignition system is unnecessary in direct-injection engines and is typically replaced with a glow-plug system for preheating air during the compression cycle.

Nonrotary combustion engines are typically classified based on their number of cylinders, the configuration of those cylinders, the air volume they displace in a single complete engine cycle, and the location of the camshaft. Camshafts are metal rods that have teardrop-shaped cams machined at certain intervals along their length (Figure D.1). Pushrod engines all have a camshaft internal to the engine block. In a pushrod engine, when the long portion of the cam reaches a rod, the rod pushes open the valves that let fuel and air into the combustion chamber and exhaust out of the engine. Overhead-cam (OHC) engines have the camshafts mounted directly atop the valves, where the cams contact the valves themselves rather than through a pushrod/rocker-arm intermediary. Engines in cars typically have between 3 and 16 cylinders, with most domestic automobile engines coming in 4-, 6-, or 8-cylinder configurations. The cylinders can be oriented perpendicular to the ground and in a line, which is called an *inline engine*. They can also be lying parallel to the ground, and are generally called *flat engines* or *boxer engines* in this configuration. It is also possible for the cylinders to be at some intermediate orientation with respect to the gravity normal (a vector perpendicular to the surface of the Earth),

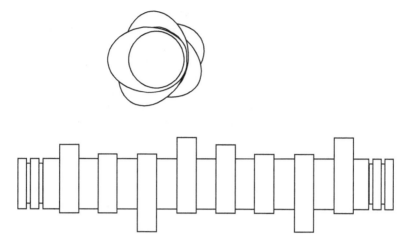

Figure D.1 Lobe design of a camshaft. Top view is along the major axis of the camshaft, while the bottom view shows the camshaft perpendicular to the major axis.

which is the case in the V- and W-style engines. Each configuration has some advantages and disadvantages, mostly associated with the physical space the engine requires, the center of gravity, and torque (since the role gravity plays in piston motion varies with the piston orientation). Thus, a 3.5-L OHC V-6 engine is a six-cylinder engine with the cylinders oriented at some fixed angle between 0° and 90° with respect to the gravity normal, an overhead camshaft, and the ability to displace a total air volume of 3.5 L when it runs through one combustion cycle.

Cars can also be classified by the location of the engine in the chassis. Front-engine vehicles have the engine mounted in the very front of the car under the hood—the most common configuration in domestic automobiles. However, there are two other places to mount an engine. Mid-engine cars have the engine mounted just behind the driver and in front of the rear wheels. Rear-engine cars have the engine mounted in the very back, sitting directly over or behind the rear wheels. Porsche 911s all have rear-engine configurations, while the Porsche Boxters are mid-engine vehicles. The choice of engine configuration in a vehicle is mainly driven by weight-balance considerations, which has a significant impact on handling (the ability of a vehicle to corner at high speed and respond quickly to a steering input from the driver).

The Wankel rotary engines are substantially different, and Mazda is the only manufacturer of domestic automobiles currently using rotary engines. In rotary engines, instead of cylinders, there are one or more teardrop-shaped devices known as rotors. The rotor is mounted to the crankshaft (called an *eccentric shaft*) mounted at the center of the engine.

The lobes of the teardrop rotor create three chambers of variable size within the rotor housing. As the rotor rotates, it takes in fuel and air, compresses that mixture, and ignites that mixture (usually with a dual-sparkplug system) to start the combustion process. The expanding combustion gases push the rotor around, and the rotor forces the expanded exhaust gases out an exhaust port. Rotary engines are still rated by a displacement equivalent to the air they displace in a single engine cycle and often by the number of rotors in the engine.

D.4   Transmissions

Transmissions are devices that help to redirect the energy harvested by the engine to the wheels, and they exert significant control over the final drive ratio (the number of revolutions of the flywheel divided by the number of revolutions of the wheels). They help the car to accelerate by allowing the engine revolutions to build quickly while the wheels are rotating slowly, taking maximum advantage of the power the engine can generate to accelerate the vehicle. Without a transmission, it would take an enormous amount of time to bring a vehicle to normal street or highway speed, if it were even possible at all. Most engines lack the torque at low rpm to accelerate a vehicle quickly. All transmissions share some common components, such as an input shaft, which is the shaft that turns with the flywheel, and an output shaft, which is mated to the input shaft via gears in the transmission and is responsible for sending power to the wheels. They also have a device that allows disengagement of the input shaft from the flywheel and a mechanism for varying the gear ratio coupling the input and output shafts. The transmission gear ratio tells you the number of revolutions of the input shaft per revolution of the output shaft, and changing this ratio is the primary method of varying the final drive ratio of the vehicle. There are two major types of transmissions found in modern consumer automobiles: the manual transmission and the automatic transmission.

Manual transmissions have several fixed gear ratios (typically four to seven) and require the driver to mechanically select the active gear ratio in the transmission. They involve both a manually actuated clutch and a manual gear selector. The clutch is a plate with springs and friction material that makes contact between the flywheel and the transmission input shaft. When it is engaged, the engine drives the transmission; when the clutch pedal is pressed by the driver, the clutch is disengaged, and the flywheel and transmission turn independently. As the driver moves the gear selector, forks in the transmission shift the position of the output shaft relative to the input shaft, forcing different gears on the input and output shafts to mesh and supply what the driver feels is the most appropriate gear ratio for the conditions. Manual transmissions offer the driver complete control: a driver can change between any two gears in a conventional manual transmission at any time. Sequential manual transmissions or sequential manual gearboxes (SMGs) are a special type of manual transmission. In contrast to a conventional manual transmission, the driver can only shift the transmission one gear up or down at a time in a SMG. However, sequential manuals eliminate the need for a manual clutch, allowing the vehicle to change gears more rapidly. This type of transmission is common on motorcycles and race cars, and it has been appearing in some high-performance domestic cars.

Like manual transmissions, most automatic transmissions have several fixed gear ratios available, but they are changed automatically as the car drives by hydraulically engaging various clutches and bands. Automatic transmissions typically contain planetary gear sets, which have an internal gear surrounded by three planetary gears held together within an outer ring gear (Figure D.2). As the clutches engage, they freeze various components of the planetary gear set, thereby changing the gear ratio between the input and output shafts. Either pressure-actuated or electronically actuated valves in a valve body (a device that routes pressurized hydraulic fluid in the transmission) route the hydraulic transmission fluid into different clutch packs, pressing them closed and freezing a component of the gear set. Automatic transmissions are mated to the engine using a device known as a torque converter in place of the clutch. The torque converter allows the transmission to decouple from the engine when the car is stopped and keeps the transmission from engaging before the engine has developed enough torque to get the car moving from a stop.

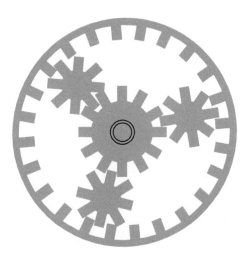

Figure D.2 Planetary gear set. The central gear is attached to the input shaft and is orbited by three planets, all contained within one outer ring gear.

Another increasingly common specialized type of automatic transmission is the continuously variable transmission (CVT), an automatic transmission that in principle can access a continuous and effectively infinite number of gear ratios between the upper and lower limits of the transmission gear ratios. In a CVT, pistons between split gears linked by a band continually vary the size of the input- and/or output-shaft gear, thereby changing the gear ratio. This type of transmission offers advantages over conventional automatic transmissions by reducing drag losses within the transmission due to rotation of nonengaged clutch packs, improving acceleration by removing time associated with gear changes, and improving fuel economy and acceleration by keeping the engine in the optimum power band for the type of driving being performed.

D.5 Suspensions

The suspension is the subsystem that supports the weight of the vehicle and permits the wheels to stay in contact with the ground while the rest of the car is free to move perpendicular to the ground, insulating the passengers from bumps and dips. The suspension typically incorporates some type of spring, either a coiled spring or longitudinal flat spring called a leaf spring, and some type of shock absorbing/damping device, typically a shock absorber or a strut. It also includes the arms, bushings, antiroll bars, and other components used to mount these devices to the vehicle and improve handling.

There are several types of suspensions. One is the fully independent suspension, in which all four of the wheels are allowed to move independent of the body and one another. Wheels on the same axle are often linked with antiroll bars to prevent side-to-side rocking motion without linking the vertical motion of the two wheels. There are a large number of configurations for attaching the suspension to the vehicle, and the interested reader is referred to other sources for a complete discussion. A second major suspension type is the live-axle suspension, where the vertical motion of both wheels on an axle are linked. Typically only the rear axle of a vehicle is a live axle, and this suspension type rarely appears on performance vehicles.

There are many types of springs and spring arrangements in a suspension. The spring in leaf-spring suspensions is composed of several narrow, bent metal bars layered on top of one another and connected to the vehicle at the bar ends and to the wheels/axles at some point in the middle of the bars. These typically are found on the rear axles of trucks and older cars. Most suspensions use the alternative coil-shaped springs, and there are two ways to incorporate a coiled spring into the suspension. In a coil-over system, the spring coil and the shock absorber are assembled as a single unit. More conventional arrangements have a separate spring and shock absorber that can be changed independently. The last common type of spring is the air-spring, or air-bag, suspension. Here, the vehicle has a pump that pressurizes air in bags that lie between the chassis and the wheels/axle. The pressurized air serves as the spring and can be used to vary the ride height of the vehicle. Air-spring suspensions are rarely factory options on consumer vehicles.

As with springs, there are many types of shock absorbers available, but only a few appear regularly in modern vehicle suspensions. Gas-filled shock absorbers have a gas-filled chamber in a piston-type arrangement, where the piston head has one-way valves. An impact compresses the gas in the chamber, which takes energy, and the energy is dissipated when the one-way valves open and allow the gas in the piston to reexpand. Fluid-filled shock absorbers work in the same way, but typically contain oil or some other liquid-state fluid. There are also electronically and magnetically controlled suspensions, where either electric or magnetic fields influence the viscosity of an electrorheological or magnetorheological fluid in the shock piston. These are becoming significantly more popular in domestic sports cars and high-end passenger vehicles.

Suspensions are complicated, and there are many designs, variables (such as the spring rate or natural stiffness of dampers), and combinations of components that generate the handling characteristics of a vehicle. A complete discussion is well beyond the objective and scope of this appendix. The interested reader is referred to engineering texts for more details.

D.6   Intake system

The intake subsystem draws air from outside the vehicle into the combustion chamber and also mixes the fuel and air in conventional gasoline internal combustion engines. The intake system must facilitate air getting to the engine quickly and with the least turbulence and restriction as possible. Air is usually pulled through some orifice open to the outside world, through a filtration system designed to remove dust and debris, and down a pipe through a valve that is controlled by the throttle called a *throttle body*. After passing through this valve, it enters plumbing that channels the air to the intake valves of the individual cylinders in a device called an *intake manifold*. Some air intakes may also include silencers, which reduce noise in the passenger cabin at the expense of introducing power-robbing turbulence into the intake system.

Most vehicles are "normally aspirated," but other forms of induction are also available. In normally aspirated engines, the movement of the pistons creates a pressure drop that naturally sucks air into the combustion chambers. Turbocharged engines and supercharged engines both use centrifugal compressors to force air into the engine at pressures greater than atmospheric and are called *forced-induction systems*.

Sometimes air-intake systems are classified by where the intake port is located. Most vehicles have what would be considered warm-air intake systems, where the intake port is somewhere in or near the engine compartment where air is warm. Cold-air intakes move the inlet port to a cooler part of the vehicle that is generally further from the engine. Cold-air intakes increase power slightly by allowing air with a greater density into the combustion chamber. Ram-air intakes put the air inlet somewhere on the front of the vehicle facing forward and with little restriction, where the forward motion of the vehicle forces more air into the engine.

D.7   Exhaust systems

The exhaust system manages the waste gases from the combustion process, removing them from the engine and processing them to reduce pollution. They effectively work on many of the same principles as the intake system: less restriction and faster gas movement equals greater power. Exhaust that exits the combustion chambers enters the system through either an exhaust manifold or a header. Exhaust manifolds simply collect all of the gas into a single pipe. Headers are designed so that the length of pipe from each cylinder to the point at which the pipes combine is identical, permitting a high and uniform exhaust velocity from each cylinder that leads to maximum removal of the exhaust gases from the combustion chamber with the least work possible. Headers typically increase the power of an engine by reducing flow restriction and differences in exhaust flow between the cylinders. From the headers/manifolds, the exhaust gases enter a collector that combines the exhaust from several cylinders together in a single pipe. After the collector, they enter the catalytic converter, a ceramic device that supports precious-metal catalysts that convert pollutants and unburned fuel into less harmful chemicals. Then the exhaust gases travel down a pipe through one or more mufflers that decrease the sound volume of the exhaust system and/or tune the exhaust to produce certain audio frequencies and volumes. After the mufflers, the exhaust enters the tailpipe and leaves the vehicle, mixing with the atmospheric gases.

D.8   Fuel systems

The purpose of the fuel system is to get the combustible fuel to the combustion chamber. In diesel engines and direct-injection gasoline engines, the fuel is injected into hot compressed gases after the intake valve is closed. Internal combustion engines mix the fuel and air before the inlet valve opens, either through fuel injection or carburetion. This discussion focuses more on internal combustion gasoline engines, since these are the more common engine type in passenger vehicles.

Carburetors are an older system for mixing the fuel and air. A carburetor controls the flow rate of air into the vehicle and relies on the suction power of the air to draw and mix fuel with the atmospheric inlet gases. Carburetors contain a venturi tube, or a tube that narrows and then widens, to accelerate the air right next to the plumbing that brings fuel into the carburetor. As that fast-moving air passes the fuel inlet port, it draws out fuel and mixes it with the air. The gas pedal controls a throttle cable that regulates a valve in the carburetor, thereby controlling the air velocity and amount of suction at any point in time. All carburetors mix the fuel and air before the air enters the intake manifold.

Fuel-injection systems are much more common in modern cars. Modern fuel-injection systems use a specialized electronically controlled injector with a very narrow nozzle to spray the appropriate amount of pressurized fuel into the inlet air. Use of the narrow nozzle atomizes the fuel, leading to rapid evaporation and improved fuel–air mixing versus carbureted systems. Fuel gets to the injector via a fuel pump, often mounted in the gas tank, that pressurizes the fuel lines and fuel rails at a constant pressure. Fuel delivery to the engine is related to the length of time the injector is open rather than pressure variations in the system. In electronic fuel-injection systems, a computer gathers data from several sensors in the intake and exhaust to calculate the appropriate fuel/air ratio and adjusts fuel delivery continuously to optimize this ratio for power or fuel economy. Fuel-injection systems come either as single-point, port, or direct-injection systems. In single-point injection, the fuel for all cylinders is sprayed into the air immediately after the throttle body, mimicking the setup of the carburetor but with improved mixing via fuel atomization. Multiport injection systems have several injectors that spray the precise amount of fuel into the inlet air directly before the intake valve. Direct-injection systems spray the fuel directly into the cylinder during the downward intake stroke.

D.9   Electrical system

The electrical system in your car provides electrical power to all the accessories in the vehicle, the ignition system (or spark system), and the starter motor that starts the engine. It includes wiring and fuses as in your home, but also the battery, alternator, starter motor, and other parts. Most internal combustion engines in passenger vehicles are started by an electric motor. When you turn the key to the ignition position, the starter motor engages with the crankshaft via a gear and draws electrical power from the car battery to rotate the crankshaft. The rotation draws in air and fuel as well as initiating operation of the ignition system. After a few rotations of the crankshaft, combustion begins, and either electronics or a turn of the key to the "run" position causes the gear in the starter motor to disengage from the crankshaft. At this point, the engine is running and the battery in the vehicle is significantly drained, since it requires a large amount of current to generate the torque required to turn the crankshaft. All engines have a device called an *alternator* mounted to the engine that turns with the crankshaft thanks to a pulley system. The alternator is essentially a mini-generator that contains a fixed coil of copper wire and a rotating magnet. As the magnet rotates in the alternator due to rotation of the engine crankshaft, electromagnetic induction causes an electric current to form that both powers the accessories while the vehicle is running and recharges the battery so that it is ready for another starting cycle.

The ignition system is the part of the electrical system that causes the spark plugs to fire in an internal combustion engine. At the appropriate time, the ignition system sends a current down a wire called a *spark plug wire* to a spark plug mounted in each cylinder. The spark plug has a pair of well-insulated electrodes separated from one another by a gap. One electrode is attached to the spark plug wire and ignition system, while the other is grounded to the engine. When current enters the spark plug, a voltage differential builds between the charged and grounded electrodes. Eventually, the voltage becomes high enough that the electrons jump the gap between electrodes, producing a spark in the combustion chamber that is hot enough to initiate the combustion reaction of the gases. Spark timing in older vehicles is controlled by a spinning device called a *rotor* that would make brief electrical contact with each spark plug wire, causing the plug to fire. Now, most cars have computer-controlled spark timing and use an ignition coil to send current to the appropriate spark plug at the appropriate time.

D.10   Braking system

The braking system is a hydraulic subsystem that is used to dissipate the kinetic energy of a vehicle, allowing it to stop. When you step on the brake pedal in a modern disk brake vehicle, you pressurize an incompressible fluid in a device known as a *master brake cylinder*. The pressure generated by this action is transmitted through the brake lines to pistons at each wheel. The pistons are housed in a device known as a *brake caliper*. The pressure causes the brake caliper to squeeze on two pieces of friction material called *brake pads*, which make contact with a metal disk known as a *rotor* that rotates with the wheels. The friction generated by this contact causes the kinetic energy to dissipate as heat, slowing the vehicle. The harder you step on the brake pedal, the greater is the squeezing force, leading to higher friction between pad and disk and faster stopping. Removing your foot from the brake pedal drops the pressure and allows the piston in the calipers to release, enabling free rotation of the wheels.

Drum brakes are also still found on the rear wheels of some cars. Drum brakes operate somewhat differently than disk brakes. In drum brakes, the friction materials, pistons, and springs are located within a hollow metal cylinder called a *drum*. The drum rotates with the wheel, and when the brake is pressed, the brake fluid forces the friction material into contact with the inside edge of the brake drum, slowing the vehicle. Drum brakes in general are less efficient and provide reduced stopping power versus disk brakes.

All cars also contain an emergency brake, which is a means of actuating the braking system mechanically. Clearly, if the pressurized hydraulic system fails while the vehicle is in operation, a nonhydraulic means of stopping the car is required. Most emergency brakes are connected to one or both of the rear brakes via a cable. Stepping on or pulling up the emergency brake lever tightens this cable and pulls the brake caliper closed, generating friction in the rear brakes and stopping the car. Emergency brakes are also used to help keep the vehicle from moving when parked for short periods of time. Leaving emergency brakes active during long periods of inactivity can cause the brakes to seize.

Cars with antilock brake systems (ABS) also contain sensors, pumps, and electronics that can rapidly pulse the brake calipers under heavy braking, preventing a very high coefficient of friction from stopping the wheel rotation suddenly. If this happens, we say the wheels are *locked* and the brakes are no longer removing energy as heat. Most ABS systems have a sensor mounted to the wheel that detects wheel rotation. If wheel rotation stops abruptly as it would during wheel lock, the ABS is activated and modulates the braking pressure many times per second. Ideally, this allows the otherwise locked wheel to rotate and the brake system to continue dissipating energy as heat. ABS systems also prevent skidding or rotation of the vehicle and permit the driver to maintain control in less-than-ideal road conditions.

Appendix E: Advanced extension exercises for capstone courses

The chemistry of cars is ideal material for capstone-type courses in chemistry and has been used successfully in this regard by the authors. There are two basic general strategies for extending car chemistry to upper-level chemists and engineers that we have used, both of which put the students in a position to work within the *synthesis* and *evaluation* levels of Bloom's learning taxonomy while drawing upon topics from potentially all of their chemistry courses. There are also many opportunities for individual instructors to exercise creativity by developing challenging questions, projects, or laboratory experiences motivated by automotive chemistry that draw upon a foundation of knowledge that crosses and extends beyond the typical undergraduate chemistry curriculum.

The first general strategy is to ask upper-level students to remove many of the assumptions built into the examples and thought exercises presented in the book. Ask the students to generate or apply a new model to a problem and quantitatively identify what additional complexities lead to a truly significant improvement of the original model. For example, challenging junior and senior chemists to rethink the turbocharger versus normally aspirated example from a van der Waals perspective is an ideal basic application of this approach. Have the students independently locate the key van der Waals parameters in the literature, determine where they appear in the calculation, and then establish the percent difference between the van der Waals and ideal-gas models. A more challenging project that extends beyond chemistry is to assemble small student groups to design and complete a life-cycle analysis to find the net energy gain/loss of various biofuels. These types of exercises force students to think creatively, explore topics from their other chemistry courses in more detail, work on individual or group general problem-solving skills, and critically evaluate their results based upon their prior experiences in chemistry and their other courses.

The second basic approach to extending car chemistry is to pose loosely framed questions that require the class to identify and fully explain how a particular chemical concept extends to another aspect of automobile production, ownership, etc., not covered in the text. This approach offers the students an opportunity to modify the course content to their particular interests while also forcing the students to form and then develop new connections, be they between chemistry and their daily lives or between chemical concepts. For example, one could envision connections between gas-law topics and the mechanisms of heat removal in brakes—vented brakes versus solid rotors or the convective role of body vents channeling air to the brakes are both areas where the basic chemistry concepts covered here can be used to understand a new function of a car. We have applied this strategy by including a semester-long project for students to research some aspect of automotive chemistry not directly covered in the text/course and to present their findings as a class lecture toward the end of the semester.

To help in identifying possible extensions for advanced courses, we have included several suggestions and linked them back to the appropriate chapter of origin.

Chapter 1: The Properties and Behavior of Gases
- Compare the pressure changes with temperature for air and pure nitrogen in a tire over a range of temperatures using the more advanced van der Waals model.
- Calculate the energy of an impact that results from a specific car hitting a speed bump or rumble strip; then calculate pressures, flow rates, etc., of gas across the dashpots in a gas shock absorber that are required to dissipate this energy in a given amount of time using the Joule–Thompson equation.
- Determine a synthetic method for producing stable magneto-rheological fluids and perhaps even test this in the laboratory.
- Find and analyze (thermodynamically and with gas laws) one of the modern air-bag chemistries.
- Challenge students to explain the role of body vents channeling air to brake rotors using the gas laws.

Chapter 2: Combustion, Energy, and the IC Engine
- Calculate the difference in horsepower produced by an engine at the start and finish of the Pike's Peak Hill Climb, where cars start at an elevation of ≈9000 ft and end at 14,110 ft. Incorporate the density changes in air based on both pressure and temperature.
- Determine the range of chemical constituents in diesel fuel and gasoline using GC or GC/MS.
- Compare the combustion by-products of gasoline versus biofuels using an IR gas cell or other analytical approach.

- Attempt a class-wide group life-cycle analysis of ethanol biofuel from different types of feedstock. Have the student groups identify the most promising approach and argue their reasoning.
- Perform detailed efficiency calculations for internal combustion (IC) engines using the Carnot, Otto, and/or Atkinson cycles. Compare aluminum and iron blocks.
- Calculate and compare the energy value of fuels using bomb-calorimetry data gathered by the students.
- Compare combustion in O_2 versus combustion in N_2O-derived O_2 in the laboratory or through Hess's law calculations.
- Mathematically calculate the effect of water or methanol injection for cooling the inlet gas temperature on energy yield or engine efficiency.
- Calculate the energetics of fatty acid methyl esters versus typical petroleum diesel molecules during combustion.
- Use cafeteria waste vegetable oil to produce biodiesel in the lab for other laboratory testing or use on campus.

Chapter 3: Oxidation and Reduction

- Design a sacrificial-anode method of protecting an automobile body from rust. Determine how often the anode will need to be changed under typical winter conditions where you live.
- Determine the optimum arrangement of batteries in a Li-ion battery pack for a Tesla Model S to get the appropriate balance of power and protection against dead cells using the data provided in the text.
- Have students experimentally determine the efficiency and chemical consequences of a catalytic converter at various temperatures in the laboratory using the natural gas lines and various analytical approaches.
- Have the students galvanize basic steel and run a corrosion experiment to determine the corrosion rates of the galvanized and nongalvanized steels.
- Run through a chrome plating process in the laboratory and have students write out the key reactions, potentials, etc.
- Delve into the details of plating chemistry using the *Modern Electroplating* text referenced in the chapter.

Chapter 4: Intermolecular Forces

- Challenge students to develop a stable emulsion of vegetable oil and water.
- Coat various metal surfaces with various types of wax or polish and measure the water contact angles. See how these vary with exposure to temperature, salt, washing with detergents, etc.
- Explore in greater detail the photochemistry of carnauba wax and its degradation.

- Have students research and present the chemistry of waterless car-wash products. Argue as a class about who would and wouldn't be willing to use such a product and why.
- Prepare various ethylene glycol/water solutions and measure the freezing-point depression or boiling-point elevation in the laboratory. See if it matches predictions based on the model calculations in the text. Make solutions of similar concentrations with other water-soluble materials and see if the solute identity really doesn't matter.

Chapter 5: Managing Heat

- Ask students to explain why the freezing point of a solvent is less sensitive to pressure than the boiling point.
- Ask students: Are there salts that would be ineffective at melting road ice? What salts are chemically the best and worst options for this and why?
- Assign groups a known refrigerant and have them design an air conditioning system, or give them the properties of an air conditioning system and have them identify appropriate refrigerants for that design. Determine the upper atmospheric temperature in which a typical automotive air conditioning system can function.
- Build a small radiator system and compare the relative abilities of air, pure nitrogen, and pure argon to remove heat from the device. Relate back to the KMT (kinetic molecular theory) properties of the gases.

Chapter 6: Materials Chemistry

- Assign student groups a polymer and challenge them to explain how it is prepared, how it is used in a car, and to bring in examples. Ask them to contact a company that produces the polymer.
- Ask the class to select a polymer that could be used to make a plastic version of a car part not typically made of plastic. Note that BMW was using plastic radiators in the mid to late 2000s.
- Vulcanize commercial latex products in the laboratory.
- Compare the ability of rubbers to incorporate various functionalized silicas.
- Ask students to build and operate a fuel cell with various hydrogen ion sources and compare the output voltage.

Chapter 7: Light and Your Car

- Provide students with some guidelines and have them use a quantum chemical calculation program to design a pigment with specific light-sorption characteristics.
- Have students look up a pigment and explain the synthesis methods, colors, and mechanisms for generating color in the pigment.

- Ask students to research the bond-dissociation quantum yields for various key functional groups or chromophores in automotive polymers, rubbers, and plastics.
- Determine the gas composition inside various automotive lightbulbs.
- Draw the energy-level diagram for various light-emitting diodes (LEDs) and calculate the emission wavelengths.

Selected bibliography

Chapter 1

Baudendistel, T. A., P. N. Hopkins, I. Linsenker, and M. L. Oliver. "Magnetorheological Fluid Damper." European Patent 1,270,988 B1, filed June 19, 2001, and issued July 27, 2005.

Chase, M. W., J. L. Curnutt, A. T. Hu, H. Prophet, A. N. Syverud, and L. C. Walker. "JANAF Thermochemical Tables." *Journal of Physical Chemistry Reference Data* 3 (1974 Supplement): 311–480.

Goldie, J. H., J. Kiley, and J. R. Oleksy. "Continuously Variable Transmissions Using Magnetorheological Fluid or Oil Shear and Methods of and Systems for Using the Same in a Vehicle, In-Wheel Application." World Patent 2,003,054,417 A2, filed Nov. 8, 2002, and issued July 3, 2003.

Lide, D. R., ed. *CRC Handbook of Chemistry and Physics*, 80th ed. Boca Raton, FL: CRC Press LLC, 1999.

Marur, P. "Magnetorheological Fluid and Method of Making the Same." World Patent 2,011,035,025, filed Sept. 16, 2010, and issued Mar. 24, 2011.

Mooney, T., G. B. Little, and G. Little. "Inflator for Vehicular Airbags." US Patent 5,713,595, filed Oct. 31, 1994, and issued Feb. 3, 1998.

Perry, R. H., and D. W. Green, eds. *Perry's Chemical Engineer's Handbook*, 7th ed. New York: McGraw-Hill, 1997.

Reji, J., and P. Narayana Das Janardhana. 2002. "A Magnetorheological Fluid Composition and Process for Preparation Thereof." World Patent 2,002,045,102 A1, filed Oct. 3, 2001, and issued June 6, 2002.

Chapter 2

Chang, R. *General Chemistry: The Essential Concepts*, 5th ed. New York: McGraw-Hill, 2008.

Collins, C. "Implementing Phytoremediation of Petroleum Hydrocarbons." In *Methods in Biotechnology*, Vol. 23, 99–108. Totowa, NJ: Humana Press, 2007.

FuelEconomy.gov. "Engine Technologies." http://www.fueleconomy.gov/Feg/tech_engine_more.shtml#dfi (accessed June 2, 2014).

Gregg, F., and C. Goodwin. *SVO: Powering Your Vehicle with Straight Vegetable Oil*. Gabriola Island, BC: New Society Publishers, 2008.

Lide, D. R., ed. *CRC Handbook of Chemistry and Physics*, 80th ed. Boca Raton, FL: CRC Press LLC, 1999.

Parker, A. "How Direct Injection Engines Work." HowStuffWorks.com. http://auto.howstuffworks.com/direct-injection-engine.htm (accessed June 2, 2014).

Perry, R. H., and D. W. Green, eds. *Perry's Chemical Engineer's Handbook*, 7th ed. New York: McGraw-Hill, 1997.

Seaton, M. Y., and G. B. Sawyer. "Varnish Analysis and Varnish Control: Molecular Weights of Vegetable Oils." In *Paint, Oil, and Drug Review* 61 (1916): 18–21.

Chapter 3

Baird, C., and M. Cann. *Environmental Chemistry*, 4th ed. New York: W. H. Freeman and Co., 2008.

Buchmann, I. *Batteries in a Portable World: A Handbook on Rechargeable Batteries for Non-Engineers*, 2nd ed. Richmond, BC: Cadex Electronics, 2001.

Carter, B. H., D. Green, and T. Hutchings. "Lubricant Method and Compositions." World Patent 1,990,005,767 A1, filed Nov. 17, 1989, and issued May 31, 1990.

Chatterjee, D., O. Deutschmann, and J. Warnatz. "Detailed Surface Reaction Mechanism in a Three-Way Catalyst." *Faraday Discussions* 119 (2001): 371–84.

Gräfen, H., E. M. Horn, H. Schlecker, and H. Schindler. "Corrosion." In *Ullmann's Encyclopedia of Industrial Chemistry*, Vol. 28, edited by B. Elvers. New York: Wiley-VCH, 2000.

Hannour, F., and Rolland, P. "Corrosion and Its Manifestation in Automotive Structures." In *Corrosion Prevention and Control*, edited by B. Raj. Oxford, UK: Alpha Science International, 2009.

Kronstein, M. "Phosphatization of Steel Surfaces and Metal-Coated Surfaces." US Patent 4,233,088, filed Mar. 29, 1979, and issued Nov. 11, 1980.

Lide, D. R., ed. *CRC Handbook of Chemistry and Physics*, 80th ed. Boca Raton, FL: CRC Press LLC, 1999.

Maclean, H. L., and L. B. Lave. "Life Cycle Assessment of Automobile/Fuel Options." *Environmental Science and Technology* 37 (2003): 5445–52.

Perry, R. H., and D. W. Green. *Perry's Chemical Engineer's Handbook*, 7th ed. New York: McGraw-Hill, 1997.

Yoder, J. A. *US Department of Energy Handbook: Primer on Lead-Acid Storage Batteries*. DOE-HDBK-1084-95, United States Department of Energy, 1995.

Chapter 4

Beckley, B. "Use the Right Citrus-Based Cleaning Product to Avoid Corrosion or Rust." United States Forest Service Technology and Development Program 0673-2319-MTDC, 2006.

Danner, B. "Amino-Functional Silicone Waxes." US Patent 7,511,165 B2, filed Oct. 4, 2004, and issued Mar. 31, 2009.

Dow Corning. "Auto Polish." *Auto Care Formulation Information*. doc 26-099B-01, Dow Corning Automotive Solutions. 2012. http://www.dowcorning.com/content/publishedlit/26-099B-01.pdf (accessed June 2, 2014).

Holde, D. *The Examination of Hydrocarbon Oils and Saponifiable Fats and Waxes*. New York: J. Wiley and Sons, 1922.

Nailen, R. L. "Grease: What It Is; How It Works." *Electrical Apparatus*, April 2004. http://www.barks.com/eaissues/04-02feat.html (accessed June 2, 2014).

Smith, K. R. "Wax Composition, Method for Manufacturing, and Method for Waxing." US Patent 8,449,663 B2, filed Aug. 31, 2005, and issued May 28, 2013.

Tuszynski, W., R. Michalczewski, W. Piekoszewski, and M. Szczerek. "Modern Automotive Gear Oils: Classification, Characteristics, Market Analysis, and Some Aspects of Lubrication." In *New Trends and Developments in Automotive Industry*, edited by M. Chiaberge. Rijeka, Croatia: InTech, 2011. http://www.intechopen.com/books/new-trends-and-developments-in-automotive-industry/modern-automotive-gear-oils-classification-characteristics-market-analysis-and-some-aspects-of-lubri (accessed June 2, 2014).

van der Linde, W. B. "Car Waxes with Improved Water-Beading Durability." US Patent 4,398,953 A, filed Oct. 26, 1981, and issued Aug. 16, 1983.

Wolstoncroft, R. L. "Method of Preparing and Packaging Automobile Wax." US Patent 4,592,934 A, filed Nov. 2, 1984, and issued June 3, 1986.

Chapter 5

Baird, C., and M. Cann. *Environmental Chemistry*, 4th ed. New York: W. H. Freeman, 2008.

Brown, T. L., H. E. LeMay, B. E., Bursten, and J. R. Burdge. *Chemistry: The Central Science*, 9th ed. Englewood Cliffs, NJ: Prentice Hall, 2002.

Delphi Thermal Systems European Headquarters. *Air Conditioning Compressors*, 2006. http://delphi.com/pdf/contact/brochures/Delphi_Compressors.pdf (accessed June 2, 2014).

Gardziella, A., L. A. Pilato, and A. Knop. *Phenolic Resins: Chemistry, Applications, Standardization, Safety and Ecology*, 2nd ed. New York: Springer, 2000.

Lide, D. R., ed. *CRC Handbook of Chemistry and Physics*, 80th ed. Boca Raton, FL: CRC Press LLC, 1999.

Perry, R. H., and D. W. Green. *Perry's Chemical Engineer's Handbook*, 7th ed. New York: McGraw-Hill, 1997.

Reisch, M. C. "Carmaker Collides with Coolant Rules." *Chemical and Engineering News* 90 (2012): 34–35.

Rugh, J., and R. Farrington. *Vehicle Ancillary Load Reduction Project Closeout Report: An Overview of the Task and a Compilation of the Research Results*. National Renewable Energy Laboratory Report TP-540-42454, 2008. http://www.nrel.gov/vehiclesandfuels/pdfs/42454.pdf (accessed June 2, 2014).

Chapter 6

Atkins, P., J. de Paula, and R. Friedman. *Quanta, Matter, and Change: A Molecular Approach to Physical Chemistry*. New York: W. H. Freeman, 2008.

Callister, W. D. *Materials Science and Engineering: An Introduction*, 5th ed. New York: John Wiley and Sons, 1999.

Dogadkin, B. A. "The mechanism of vulcanization and the action of accelerators." *Journal of Polymer Science* 30, no. 121 (1958): 351–61.

Grot, W. G., C. J. Molnar, and P. R. Resnick. "Fluorinated Ion Exchange Polymer Containing Carboxylic Groups, Process for Making Same, and Film and Membrane Thereof." US Patent 4,544,458 A, filed June 16, 1980, and issued Oct. 1, 1985.

Grulke, E. A. *Polymer Process Engineering*. Lexington: TPS Publishing, 1994.

Gupta, V. B., and V. K. Kothari. *Manufactured Fibre Technology*. London: Chapman and Hall, 1997.

Hertz, D. L. "Theory and Practice of Vulcanization." *Elastomerics* Nov. 1984. http://www.getallpdf.net/files/vulcan.html (accessed June 2, 2014).

Karabin, S. "Making Tires 'Green.'" *Chemmatters* 2013. http://www.acs.org/content/acs/en/education/resources/highschool/chemmatters/news/making-tires-green.html (accessed June 2, 2014).

Kato, A., A-P. Tsia, and M. Watanabe. "Magnesium Alloy." European Patent 1,813,689 A1, filed Sept. 21, 2005, and issued Aug. 1, 2007.

Keller, U., and H. Mortelmans. "Adhesion in Laminated Safety Glass: What Makes It Work?" In *Proceedings of the 6th International Glass Conference*, Tampere, Finland, 1999, 353–56.

Kogel, J. E., N. C. Trivedi, and J. M. Barker, eds. *Industrial Minerals and Rocks: Commodities, Markets, and Users*. Englewood, CO: Society for Mining, Metallurgy, and Exploration, 2006.

May, C. A. *Epoxy Resins: Chemistry and Technology*, 2nd ed. New York: Marcel Dekker, 1987.

Puppin, G. "Thermoplastic Resin and Fiberglass Fabric Composite and Method." EU patent 0,867,270 B1, filed Nov. 19, 1997, and issued Apr. 3, 2002.

Schoenlaub, R. A. "Glass Composition." US Patent 2,334,961, filed Dec. 5, 1940, and issued Nov. 23, 1943.

Seeboth, N., S. Ivanov, and S. Molkov. "1,2,4-Triazine Suitable as a Vulcanization Accelerator and Method for Producing Same." US Patent 20,130,053,559 A1, filed Oct. 8, 2010, and issued Feb. 28, 2013.

Tiede, R. L., and F. V. Tooly. "Glass Composition." US Patent 2,571,074, filed Nov. 2, 1948, and issued Oct. 9, 1951.

Tullo, A. H. 2009. "Stretching Tires' Magic Triangle." *Chemical and Engineering News* 87, no. 46 (2009): 10–14.

Wipke, K., S. Sprik, J. Kurtz, T. Ramsden, C. Ainscough, and G. Saur. *National Fuel Cell Electric Vehicle Learning Demonstration Final Report*. National Renewable Energy Laboratory technical report NREL/TP-5600-54680, 2012. http://www1.eere.energy.gov/hydrogenandfuelcells/pdfs/learning_demo_final_report.pdf (accessed June 2, 2014).

Chapter 7

Berkei, M., U. Nolte, and T. Sawtowski. "Stabilization of Organic Polymers against Free Radicals." US Patent 8,410,206, filed Mar. 8, 2007, and issued Apr. 2, 2013.

Collins, C. B., and E. G. Zubler. "Iodine Cycle Incandescent Lamps." US Patent 3,132,278, filed Sept. 18, 1961, and issued May 5, 1964.

Flesch, P. *Lights and Light Sources: High Intensity Discharge Lamps*. New York: Springer, 2006.

Handa, J., H. Ito, H. Hattori, and A. Suganuma. "Metallic Paint Film." European Patent 313,280, filed Oct. 14, 1988, and issued May 22, 1991.

Hebbinghaus, G., and P. Postma. "High Intensity Discharge Lamp." World Patent 2,006,043,191 A1, filed Oct. 10, 2005, and issued Apr. 27, 2006.

Horiuchi, M., T. Saito, K. Takahashi, and M. Takeda. "Mercury-Free Metal Halide Lamps." European Patent 1,037,258 A1, filed Feb. 17, 1999, and issued Sept. 20, 2000.

Kasap, S., and P. Capper, eds. *Springer Handbook of Electronic and Photonic Materials.* New York: Springer, 2006.

Lide, D. R., ed. *CRC Handbook of Chemistry and Physics*, 80th ed. Boca Raton, FL: CRC Press LLC, 1999.

Poth, U. *Automotive Coatings Formulation: Chemistry, Physics, and Practices.* Berlin: Vincentz Network GmbH & Co KG, 2008.

Smith, H. M. *High Performance Pigments.* New York: John Wiley and Sons, 2002.

Stachura, S., V. J. Desiderio, and J. Allison. "Identification of Organic Pigments in Automotive Coatings Using Laser Desorption Mass Spectrometry." *Forensic Science* 52, no. 3 (2007): 595–603.

van de Streek, J., J. Bruning, S. N. Ivashevskaya, M. Ermrich, E. F. Paulus, M. Bolte, and M. U. Schmidt. "Structures of Six Industrial Benzimidazolone Pigments from Laboratory Powder Diffraction Data." *Acta Crystallographica* B65 (2009): 200–11.

Appendix C

DeRudder, J. L. "Commercial Applications of Polycarbonates." In *Handbook of Polycarbonate Science and Technology*, edited by D. G. LeGrand and J. T. Bendler, 303–16. New York: Marcel Dekker, 2000.

FDA. "Update on Bisphenol A for Use in Food Contact Applications." 2010. http://www.fda.gov/downloads/NewsEvents/PublicHealthFocus/UCM197778.pdf (accessed June 2, 2014).

FDA. "FDA Continues to Study BPA." FDA Consumer Health Information. 2012. http://www.fda.gov/downloads/ForConsumers/ConsumerUpdates/UCM297971.pdf (accessed June 2, 2014).

Reisch, M. "What's That Stuff? Spandex." *Chemical and Engineering News* 77 (1999): 7.

Rubin, B. "Bisphenol A: An Endocrine Disruptor with Widespread Exposure and Multiple Effects." *J. Steroid Biochem. Mol. Bio.* 127 (2011): 27–34.

Serini, V. "Polycarbonates." In *Ullmann's Encyclopedia of Industrial Chemistry*, Vol. 28, edited by B. Elvers, 603–12. New York: Wiley-VCH, 2000.

Szycher, M. "Introduction." In *Szycher's Handbook of Polyurethanes*, 1–12. Boca Raton, FL: Taylor and Francis Group, 2013.

Titow, W. V. "Introduction." In *PVC Technology*, edited by W. V. Titow, 1–36. New York: Elsevier Science Publishing, 1986.

Wunsch, J. R. *Polystyrene: Synthesis, Production and Applications.* Akron, OH: Smithers RAPRA, 2000.

Index

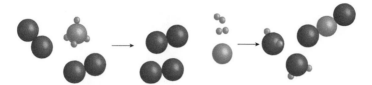

Figure 2.5 Atom-scale view of the combustion of methane from the enthalpy-of-formation perspective. From a theoretical perspective, the reactants are viewed as breaking apart into their component elements; then the elements recombine to form new product compounds.

Figure 2.8 Photo of the engine bay in a 2005 Subaru Outback 2.5 XT, highlighting the turbocharger (green oval) and the intercooler (red oval). The turbocharger is mostly hidden under a heat shield, but you can clearly see the turbo outlet bolted to the intake port of the intercooler (left side of the intercooler).

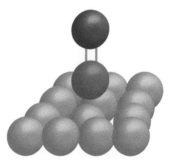

Figure 3.3 Physisorption of carbon monoxide at the surface of a platinum metal. The dark sphere represents carbon, the red sphere oxygen, and the light gray spheres the atoms of platinum at the platinum cluster surface. Note the lack of chemical bonds between the carbon and the surface.

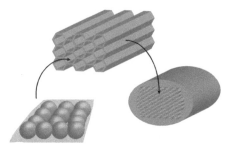

Figure 3.4 Platinum and other transition metal catalyst particles are embedded at the surfaces of a honeycomb-like ceramic support in the catalytic converter.

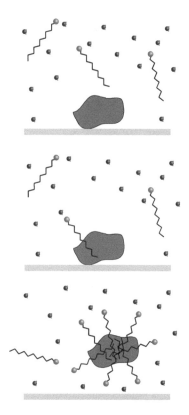

Figure 4.2 Surfactant action with an organic contaminant. In the top image, we see the water far away from the oily droplet and surfactant molecules in solution. In the middle image, surfactants begin to dissolve their hydrophobic ends in the oil droplet, bringing some water closer to the dirt. In the final image, the droplet is coated in surfactant molecules, making it look water soluble and allowing it to lift from the surface into the water solvent.

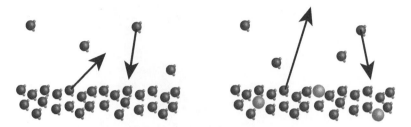

Figure 5.1 The pure solvent (left) and solution (right) liquid–vapor interfaces. The length of the arrows depicts the energy needed to change phases and shows that the solute upsets the balance between the energies, meaning that the temperature must be increased to reestablish equilibrium.

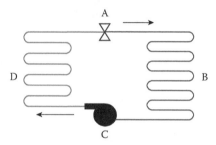

Figure 5.4 Schematic diagram of the typical automotive air-conditioning refrigeration system. Just to the right/blue side of the evaporator valve (A), the coolant is a cold liquid at low pressure. It vaporizes by removing heat from the cabin air in the evaporator coils (B), then enters the compressor (C) as a cold vapor at low pressure. When it exits the compressor, it is a hot vapor that passes through the heat exchanger (D), giving off heat to the atmosphere and condensing into a hot liquid at high pressure.

Figure 6.4 Photo of aluminum wheel corrosion underneath a polymer wheel coating.

Figure 7.1 White light with all photon energies strikes the car, but only the red photons are reflected toward your eyes. The light waves in the image correspond to blue (shortest wavelength), green (intermediate wavelength), and red (longest wavelength).

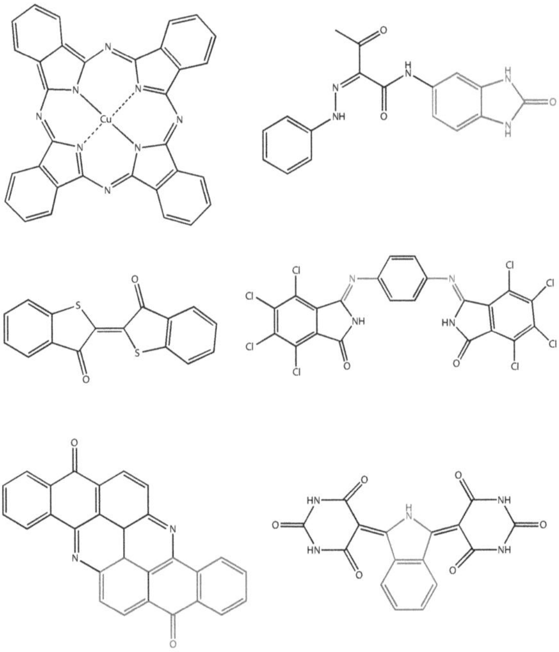

Figure 7.2 Typical categories of organic pigment used in the automotive industry. From left to right, starting in the top row: phthalocyanine blue, a benzimidazolone, thioindigo, a green azomethine, flavanthrone yellow, and a red–yellow isoindoline. Since many of these are chromophore families, the critical functional group for which the family is named is highlighted in red if the chromophore family is not variations on the entire structure. For flavanthrone, the anthrone group is highlighted.